Relearning Mathematics

Volume 1

Why do so many people have problems with maths? How can teachers help students deal with 'maths anxiety'?

Intended mainly for adult students who have been made to feel a failure at maths, this textbook overcomes learning obstacles by developing methods that help empower students. Using examples from other disciplines, such as politics and history, the text encourages students to reflect on their own learning, to analyse their work for error patterns, and to look for hidden messages in maths exercises. They learn to approach problems more carefully by first understanding, comparing and rounding numbers, before solving any arithmetical problems.

At a time when government demands are narrowing educational approaches, this book provides an alternative – a critical maths 'literacy'. It sets maths problems in the context of current issues around race, gender and class.

Marilyn Frankenstein is currently Chair of the Department of Applied Language and Mathematics, College of Public and Community Service, University of Massachusetts, Boston.

Delicate =	A 00.59 GR	= Cruel	
rain =	A 00.42 GR	= war	
dash'd =	A 00.36 GR	= wage.	
to a' =	A 00.36 GR	= Maim	
eager =	A 00.36 GR	= a naked	
field =	A 00.36 GR	= child.	
lyrically =	A 01.17 GR	= Senseless	
transcending =	A 01.28 GR	= and profound	
the bare =	A 00.59 PR	= dying.	
infirm =	A 00.69 GR	= Guilt	
boughs =	A 00.72 PR	= exacts	
to leaf =	A 00.59 PR	= a dismal	
+ twig. =	A 00.59 PR	= toll.	
808	A 08.08 TL	808	
	A 08.08		
	A 00.00		

NM YOU

2505 11 JAN 72

Frontispiece

Relearning Mathematics

A Different Third R – Radical Math(s)

Volume 1

MARILYN FRANKENSTEIN

With forewords by Henry A. Giroux and Europe Singh
and an afterword by Ubiratan D'Ambrosio

'. . . an association in which the free development of each
is the condition of the free development of all'

Free Association Books / London / 1989

First published in 1989 by
Free Association Books
26 Freegrove Road
London N7 9RQ

British Library Cataloguing in Publication Data
Frankenstein, Marilyn
Relearning mathematics : a different third R-radical
maths(s).
1. Mathematics
I. Title
510

ISBN 1-85343-091-9

Typeset by MC Typeset Ltd, Gillingham, Kent.
Printed and bound in Great Britain by Short Run Press Ltd, Exeter

Cover illustration:
Bakuba strip patterns (see ornamental end papers)
assembled from units by Claudia Zaslavsky © 1975
from an idea by Marilyn Frankenstein

Contents

A Radical Mathematics
Foreword by Henry A. Giroux

During the last two decades educational critics have made a number of important advances in developing a viable critical theory of education. In particular, radical theorists have made significant inroads in providing a language of critique by which to analyse and demystify the role that schools play as agencies of moral and political regulation. They have also begun to provide a programmatic language by which to understand schools as sites of critical learning and social empowerment.

Central to this project has been the more recent work of theorizing schooling as a form of cultural politics. In this view, the relationship between knowledge and power is analysed as part of a wider effort to define schools as places where a sense of identity, worth, and possibility is organized through the interaction among teachers, students, and texts. In this view, schools are places where students are introduced to particular ways of life, where subjectivities are produced, and needs are constructed and delegitimated. The more recent work in radical theories of education has focused on two general lines of inquiry. In the first and most dominant mode of inquiry, radical theorists have analysed the various ways in which knowledge and power come together to give a particular ideological bent to the form and content of curriculum knowledge. Much of this work is concerned with uncovering the ideological interests at work in the content of the curriculum, that is, revealing how racist, sexist, and class-specific messages work to construct particular ideological representations and images. Equally important, but to a lesser degree, radical theorists within this perspective have attempted to analyse the structuring principles of curricula texts in order to understand more fully how these coding structures contribute to the ways in which knowledge is produced, mediated, consumed, and transformed as part of the overall pedagogical process. In the second mode of inquiry, radical theorists have focused on the historical and cultural practices of subordinate groups and the ways in which these give rise to particular relations of oppression and resistance in schools. In this perspective, a great deal of attention is given to analysing how school as a cultural and social terrain organizes, legitimates, sustains, and refuses particular forms of student experience. In some cases, there have been attempts within this perspective to develop the rudiments of a critical pedagogy based on an attempt to legitimate and incorporate the everyday experiences, languages, histories, and values of subordinate groups into the school curricula. Central to this perspective is the need to view schools as both instructional and cultural sites, that is, as places where knowledge and learning are deeply related to the different social and cultural forms that shape how students understand and respond to classroom work.

As important as this approach has been it has not managed to integrate attempts to develop theoretically and politically useful school knowledge with a similar concern for developing a critical pedagogy. On the contrary, radical theorists who focus on developing 'ideologically correct' school knowledge often assume that questions of pedagogy can be treated as an afterthought, that is, they often believe that if teachers present the 'right knowledge' to students they will automatically learn something. In this case the ideological correctness of one's position appears to be the primary determining factor in assessing the production of knowledge and exchange that occurs between students and teachers. At best, questions of pedagogy are reduced to a technical consideration over whether one might use a seminar, lecture, or multimedia format. On the other hand, a number of theorists struggling with the difficult task of creating the broader outlines of what constitutes a critical pedagogy seem impervious to the issue of what knowledge might be of most worth as part of the critical reconstruction of school curricula. It is to Marilyn Frankenstein's credit that she has produced a text in which she theoretically defends integrating these positions as well as demonstrating what such an integration might look like.

What is most remarkable about Frankenstein's approach is that she situates the teaching of mathematics within a rationale that links schooling to the wider considerations of citizenship and social responsibility. For Frankenstein, to teach is to assume the position of an advocate, to believe that schools count for something and, in this case, to argue that the first responsibility of teaching is to educate for informed and active citizenship. By establishing a sense of democratic purpose and a moral and political referent for her work, Frankenstein manages brilliantly to overcome many of the polarizing dichotomies that separate radical theories of ideology and knowledge production from the language and practice of critical pedagogy. On the one hand, Frankenstein provides a theoretical and political foundation for refusing to legitimate her approach to teaching in either an exclusive appeal to the importance of a specific form of knowledge (the canon) or in an appeal to a specific set of social relations (pedagogy). Instead, she defends and develops her views on the importance of mathematics and student learning as part of a wider project of pedagogical and political transformation. On the other hand, Frankenstein removes mathematics from its hallowed place as an unchangeable part of the established canon; from its place in the culture of scientism, she appropriates it for the tradition of critical pedagogy.

For Frankenstein, maths can only be understood as a historical, social, and political construction. It is this important insight that allows her to recognize that knowledge is not merely produced in books or in the heads of teachers; nor is, in this case, mathematical knowledge an objective, scientifically sanctioned body of discrete rules, information, and laws. On the contrary, knowledge, identities, and values are produced as part of a wider pedagogical process that characterizes the interaction among teachers, texts, and students. Frankenstein demonstrates this position by combining what can be called a pedagogy of knowledge with a knowledge of pedagogy.

A pedagogy of knowledge focuses on the processes that give rise to a particular notion of a subject, idea, or discipline. In this case, Frankenstein demonstrates in her text how 'maths' as a socially constructed body of knowledge becomes part of the received canon through a variety of ideological constructions that mediate how people both perceive and appropriate it as an objective, universalized, scientific, and depoliticized discourse. Frankenstein not only demystifies the pedagogy of maths by exposing its underlying ideological constructs, she also provides students with a different pedagogical construction; that is, she restructures and reconstructs the pedagogy of maths so it can be seen and critically appropriated as part of a wider struggle for self and social empowerment. For instance, she focuses on how people come to learn 'maths anxiety', she investigates the community of memories that students bring to the classroom regarding their maths experiences, and she encourages students to treat maths as a discourse or text which contains a number of latent and obvious assumptions that need to be named and critically interrogated. As part of Frankenstein's critical pedagogy, learning maths becomes part of a broader pedagogical process of learning about how a given set of ideological and material relations position and structure our relations to each other and the dominant society.

While a pedagogy of knowledge focuses on how a body of knowledge is socially constructed as part of a learning process with a particular ideological intent, a knowledge of pedagogy points to how students produce meaning through the categories and experiences that mediate how they perceive what is presented to them as schooled knowledge. For Frankenstein, learning about how maths is constructed only becomes meaningful if it is accompanied by a pedagogy that raises questions about how it is that students produce meaning, how they come to make affective and ideological investments in particular learning situations, and how a pedagogy can be constructed in which teachers and students can learn together as part of a wider political and moral vision. Frankenstein succeeds admirably in combining these two pedagogies in her text. She brilliantly integrates into her attempts to demystify and reconstruct knowledge about maths a critical pedagogy that draws attention to the knowledge, needs, and interests that constitute the individual and collective voices of her students. For example, Frankenstein argues that knowledge of basic numerical and statistical concepts is important not merely as part of what is to be learned for a test or for understanding the logic of maths, but as part of the struggle to make the individual and collective lives of students meaningful.

Learning maths serves as a form of ideology-critique, as a way of using knowledge to undermine common-sense assumptions about how society is structured, and to act from more informed choices about the structures and processes of society. It is important to stress that Frankenstein is not merely concerned about getting students to think critically, she also encourages students to make commitments, to develop a capacity for empathy, to engage in new and more human learning communities, and to believe that they can make a difference in changing the conditions of suffering they confront in the wider society. For Frankenstein teaching maths means locating the nature of her subject-matter within the historical traditions, languages, concerns, and communities of memories that both legitimate and challenge the identities and experiences of her students. Learning for Frankenstein means breathing into the subject-matter of schooling a sense of the constraining and enabling dimensions of everyday life, it means experiencing relations of possibility, and it means understanding difference in a morally affirming way that promotes rather than undermines the possibility for democratic life.

Marilyn Frankenstein has combined the best work of radical theories of schooling with the insights currently being developed in the field of critical pedagogy, and in doing so has produced a text that will make existing and future generations of teachers rethink the basis of the relationship between knowledge and critical pedagogy. Frankenstein deftly demonstrates in clear and accessible language that learning maths is a form of literacy that cannot be removed from the larger processes of reading and writing, or from the problems of society. Her notion of mathematical literacy is about more than the ability to calculate, it is more fundamentally about the ability to understand what numbers mean – how the number system is constructed as part of a wider signifying system that actively constructs and legitimates ideologies about everyday life and the larger society. As developed in this text, mathematical literacy is about understanding both the mathematical disempowerment of people and the possibility for empowerment through maths literacy as a form of cultural politics. This is a path-breaking book and should be read by everyone concerned with politics, education, and critical pedagogy.

The British Context

Foreword by Europe Singh

In my adult-numeracy classes I am confronted with students who have used mathematics all their working lives yet who, once inside a classroom, abandon all their experience and self-perfected rules of thumb. They readily submit to the 'paper-and-pencil' they had learned at school, where they had been made to feel failures at maths. All too often, mathematics lessons devalue self-taught methods, in favour of an oppressive standardization known as 'the proper method'. This approach becomes particularly absurd when imposed upon students from non-Western cultures who may have already learned a different algorithm, a different method of calculation. Furthermore, the usual battery of individual timed tests has the effect of atomizing students, leading them to shield answers from the anxious, prying gaze of others. With such practices, it should be no surprise that most students end up feeling they are 'no good at maths'.

Why do so many fail maths at school? How do people really learn maths, as it is used in everyday life? Sharp disagreements have arisen over such questions in the long-running debate over Britain's national core curriculum for maths. Education Secretary Kenneth Baker has insisted upon reverting to 'traditional pencil-and-paper practice' and regular drill (in basic number work, measurement, algebra), as well as devising 'a clear structure of age-related targets' for test results, from as early as age seven.

In contrast, even Baker's own working group of maths educationists has warned that the difficulty of maths as an abstract subject 'cannot be overcome by repeated practice in the manipulation of symbols by pupils who do not understand their meaning'. Moreover, they said, 'The essence of problem-solving is in some sense unfamiliar and non-routine to the solver.' Accordingly, the group proposed that the new curriculum should:

- encourage students to use electronic calculators, as an alternative to traditional arithmetic drill;
- ground maths pedagogy in solving real-life problems, so that students learn to choose the right technique for a particular problem;
- monitor students' personal qualities such as motivation, perseverance and team-work, as a guide to the pedagogy's success;
- set only general targets for students, not according to age.

Bowing to governmental pressure, the National Curriculum Council has cut across those recommendations by largely adopting a return to 'basics'. Along with related changes in science education, this move involves several political agendas. A pedagogy linked to national standardized testing would assist the government's strategy of creating a 'market' in education, as schools are thrown into competition with each other for test results – and so for students, teachers and ultimately State funding. It also aims to avoid a predicted shortage of scientists, technicians and engineers in the 1990s.

Notwithstanding the government's declared aim of making maths and science more accessible to all, the curriculum changes would favour a small minority of students selected out early as the nation's future technical experts.

The changes have been supported as the remedy for progressive educational methods supposedly responsible for undermining the population's mathematics skills. However, mass numeracy has never been achieved by traditional methods, as was pointed out in the Cockroft report (1982), which reminded us of similar lamentations in Britain's history:

> In arithmetic, I regret to say worse results than ever before have been obtained . . . failures are almost invariably traceable to radically imperfect teaching.
>
> (Her Majesty's Inspectorate report, 1876)

> It has been said . . . that accuracy in the manipulation of figures does not reach the same standard which was reached twenty years ago.
>
> (a Board of Education report, 1925)

> The standard of mathematical ability of entrants to trade courses is often very low . . . there is yet no evidence of marked change.
>
> (Mathematical Association report, 1954)

The Cockroft Committee criticized maths education for offering an 'inadequate opportunity to see the skills emerging from the solution of problems'. It emphasized the need to overcome the problem of abstraction:

> Mathematics lessons in secondary schools are very often not about anything. You collect like terms, or learn the laws of indices, with no perception why anyone needs to do such things. There is an excessive preoccupation with a sequence of skills and quite inadequate opportunity

> to see the skills emerging from the solution of problems. As a consequence of this approach, school mathematics contains very little coincidental information. A French lesson may well contain some incidental information about France – so across the curriculum; but in mathematics the incidental information which one might expect (current exchange and interest rates, general knowledge on climate, games, social statistics) is rarely there.

Research on gender and mathematics (for example, Burton, 1986) has criticized maths teaching for such abstraction, as well as for inherent biases, which together serve to alienate working-class students, black students and girls. Research on 'maths anxiety' (Buxton, 1981) and 'mathophobia' has attributed these student reactions to the élitism of mathematics, continually reinforced by normative testing and a relentless diet of sums abstracted from reality, which impede students' learning from infancy through adulthood.

Ignoring all these warnings from educationists, the government attributes 'falling standards' to those who have gone furthest in making maths education more accessible. This view was expressed most crudely by the Prime Minister herself at the 1987 Conservative Party conference: 'Children who need to count and multiply are being taught anti-racist maths, whatever that might be'; according to her, they are being cheated of a sound start in life by 'hard-left education authorities and extremist teachers'. Backed up by 'New Right' ideologues, such attacks aim to legitimate the government's project of intensifying the élitist character of education, while excluding the progressive innovations of recent years. The current version of a national curriculum, with regular testing of 'basic skills', would serve to perpetuate and strengthen all the main obstacles to maths learning, and would lead to more rigid hierarchies of 'maths ability'. All this would further deter working-class, black and female students from pursuing maths and physical sciences. The majority of students would be sacrificed for the possible benefit of a few. In this manoeuvre, 'standards' are being defined narrowly, in a way that favours white middle-class students. Yet the real problem is not 'falling standards'. It is not that most students fail maths, but that school maths 'fails' the majority of students who pass through it – and has always done so.

Marilyn Frankenstein begins from that premise. She believes that the powerful tools of mathematical analysis are denied to the majority through élitism, mystification and the teaching of irrelevances. Taking further the criticisms made by other educationists, she believes that this denial serves a political purpose, as it helps to maintain exploitation and oppression in this society. Furthermore, she believes that the way society is organized is taken for granted in many attempts to make maths more relevant to students. Exactly how maths is made more relevant affects how students learn it and whether they imagine using it for formulating their own questions about the world. One way or another, maths pedagogy is unavoidably political.

How, then, do we offer a chance to relearn maths to those who have been failed by the system? How do we overcome the stultifying effects of maths schooling? Marilyn Frankenstein's unique book attempts to help empower students, in a dual way. First, it helps students to unblock their fear of maths and to recognize their own abilities to do it. Second, she places maths problems in real-life contexts that sharpen students' understanding of society and their role in it.

In all these ways, the book helps adults to relearn maths. Rather than teaching maths as something to be reproduced in tests for the teacher, the book encourages students to make maths part of their armoury of skills for interpreting and acting on the world.

Editor's Note

Math or maths? This difference in terminology is only one of many differences between US and UK usages. We have generally used the British term 'maths' in the book because it has been produced in the UK. Wherever national differences in terminology express differences in meaning, we have explained them, as with 'black' versus 'people of colour'.

It should be noted that Marilyn Frankenstein wrote an earlier version of this book for a US audience, before Free Association Books knew of it. Given the successful reception of our book *Anti-Racist Science Teaching*, and our contact with networks of maths and science teachers, we decided to publish the book with extra material from the UK. Thanks to Europe Singh's help and Marilyn's additional work, we are proud to publish our first student textbook, a unique tool for critical mathematics learning in the USA, the UK and elsewhere. Both we and the author would appreciate receiving any comments or additional examples from readers, particularly for preparing Volume 2.

L.L.

Acknowledgements

I have written this book over a ten-year period and view it (together with Volume 2) as a culmination of the teaching and learning I have done over these first twenty years of my professional work. Because of this, and because I have followed my father's example of making many friends, it has been hard to write these acknowledgements. I keep adding names; I keep making new friends who influence my work.

Some friends have been important intellectual and political mentors: Henry Giroux, James Jennings and Ubi D'Ambrosio. Also, I have been (and will continue to be) collaborating with many others on education and political projects over at least the next twenty years of my professional work: Arthur B. Powell, Claudia Zaslavsky, John Volmink, Munir Fasheh, Arthur Conquest, Bill Crombie, Frank Davis, Víctor López-Tosado, Marta Bolívar, Paulus Gerdes, Femi Oguntosin, and Ellen Banks (a brilliant artist about whose work I hope to write, one day, when I take a break from mathematics education). Still others, some of whom I've never met, supported my ideas in my tenure review, and so gave me the opportunity to continue my work: Delia Aguilar, Michael Apple, Stanley Aronowitz, Barbara Barnes, Ann Berlak, Blanca Facundo, Mary Gray, Carl Hedman, David Henderson (who also introduced me to many of my future collaborators) Richard Katz, Paul Monsky, Richard Ohmann, Anthony Petrosky, Gloria Rakovic, Danielle Rice, Hugo Rossi, David Rubin, Ira Shor, Patricia Sipe, Julian Weissglass and Joan Witaker.

The conditions of my daily work environment make it possible for me to be involved in exciting intellectual work without 'burning up'. The College of Public and Community Service (University of Massachusetts, Boston's Downtown Centre) is an institution with a strong commitment to educating urban adults who had not been 'tracked' or 'streamed' for higher education. I have learned a lot and laughed a lot with my colleagues/friends there and with my colleagues/friends on the editorial board of *Radical Teacher* magazine. My students, above all, have encouraged me to continue with the different ways and focus of my teaching, in particular Darleen Bonislowski, Andrea Booker, Lougenia Bowman, Gloria Cardona, Michelle Cole, Florence Cook, Vangie Dupigney, Mary Edwards, Loreta Fortini, Vevienne Jackson, Betty Ann Jones, Virginia Lee, Maria Simpson, Sylvia Smith, Gilberto Sosa, and Judy Tassell, as well as Shafi Abdul-Mani and Clara Laboy, two students from Park East alternative high school in East Harlem, New York City (whom I taught in the early 1970s and who are now adult friends).

Many people have helped get this book into print. Judy Ruben, at Monthly Review Press, was the first editor to express strong interest in this text. Bob Young at Free Association Books seriously followed through when, on a visit to London, I telephoned with a vague inquiry about this textbook. Europe Singh, a radical mathematics teacher in London, helped make the book more appropriate for students in Britain. Karen Shaw, a wonderful artist, contributed her fascinating work for major illustrations in the text. Anne Smith, with help from Susan Goldstein and Christine Parsons, typed the manuscript. (I am probably the only mathematics teacher in the USA or Britain who does not use a word processor!) And Les Levidow, my editor at Free Association Books, has painstakingly followed up on every detail, making many valuable suggestions and thought-provoking comments.

Finally, so many wonderful friends and their families have helped keep me alive and fighting; I could never acknowledge them all. In addition to everyone already listed I must mention: Betty Boyd, Chris Boyd, Ellen Cantarow, Connie Chan, Gina Conquest, Carmen Dillon, Richard Frankenstein (my brother), Louise Forsyth, Reebe Garofalo, Jim Green, George Gulick, Betty Johnson, Diane Jones, Louis Kampf, Kelsey Kauffman, Marie Kennedy, Tom Lally, Sharyn Lowenstein, Ruthi Massey, Charlene Morse, Laura Nemeyer (my therapist), Martha Nencioli, Billy Novick (my music teacher), Susan O'Malley, Ashanti Pasha, Nancy Phillips, Miriam Pope, Dung Quach, Glenn Rothfeld (my acupuncture doctor), Susan Rubin, Ed Rubin, Saul Slapikoff, Emilie Steele, Bruce Stinebrickner, Chris Tilly, Kathleen Weiler and Ann Withorn.

For permission to reproduce illustrations, we should like to thank the following:

Ornamental End Papers: G. Williams, *African Designs from Traditional Sources*, Mineola, NY: Dover Art, 1971; Frontispiece: Karen Shaw; p. 2: Bill Plympton; p. 15: M.C. Escher, 'Liberation'/Collection Haags Gemeentemuseum, The Hague; p. 29: Oxfam; p. 32: Karen Shaw; p. 38: Clifton Fadiman (ed.)/*Saturday Review*/Omni Publications International; p. 73: Karen Shaw; p. 82: Estelle Carol and Bob Simpson; p. 83: from Georges Ifrah, *One to Zero*, copyright © 1981 by Editions Seghers, Paris. English translation copyright © 1985 by Viking Penguin Inc. All rights reserved. Reprinted by permission of Viking

Penguin Inc.; p. 90: Jasper Johns Collection/Museum of Modern Art/Mr and Mrs Armand P. Bartos (donors); p. 103: Dana Summers, © 1988 Washington Post Writers Group/*Orlando Sentinel*, reprinted with permission; pp. 109–12: Cecilio Garcia-Camarillo; p. 122 Dan Wasserman, © 1988 *Boston Globe*, reprinted by permission of Los Angeles Times Syndicate; p. 124 Wayne Stayskal/*Chicago Tribune*, reprinted by permission, Tribune Media Services; p. 127: Hereth/Liberation News Service/Community Press Features; p. 134: Dan Wasserman, © 1986 *Boston Globe*, reprinted by permission of Los Angeles Times Syndicate; p. 135: Tony AUTH © 1985 *Philadelphia Inquirer*. Reprinted with permission of Universal Press Syndicate. All rights reserved; p. 135: Dan Wasserman © 1982 Los Angeles Times Syndicate; p. 150: Martin Adler Levick; p. 154: World Development Report; p. 156: *New Society*; p. 174: National Science Foundation; p. 177: *Progressive Agenda*, August 1986; p. 199: *In These Times*; p. 199: Tony Garrett/*New Society*; p. 203: *New Society*; p. 214: Max Bell, School Mathematics Study Group; p. 220: Laguna Sales; p. 221: *Science for the People*; p. 222: from David Sanders, *The Struggle for Health*, Macmillan; p. 222: *New Internationalist*; p. 229: *The Mathematics Teacher*; p. 253: Don Wright, *Miami News*; p. 261: Estelle Carol and Bob Simpson.

For granting us permission to reproduce extracts from their texts, we should like to thank the following: M. and R. Ascher, 'Ethnomathematics'; Philippa Braidwood, 'Money for Jam – But None for Lentils', © *Observer*; Dorothy Buerk; Cristian Cabrera, 'Pepe's Pineapple'; Munir Fasheh, 'When Structures Fall, People Rise'; Federation of University Employees (425 College Street, New Haven, CT 06511), 'A Report to the Community'; P. Freire and D. Macedo, *Literacy*; Jerome Karabel and David Karen, 'Color on the Court', *In These Times*; Ann Landers, 'A Plea to End Nuclear Threat', *Boston Globe*/Los Angeles Times Syndicate; 'Metrication at the Crossroads' and '299,792,458 Meters per Second' © 1979, 1982, Scientific American, Inc., all rights reserved; P.T. Nash, 'Why Not "Oriental"?', *Nichibei*; 'Old Values, Anyone?' reprinted by permission from *The Progressive* (409 E. Main Street, Madison, WI 53703), © 1981, The Progressive, Inc.; Oxfam, 'The Facts of Life'; Linda Pastan/Ernie Robson, 'Algebra'; J.A. Paulos, 'From One to Zero', © 1985 New York Times Company, reprinted with permission; Radical Statistics Education Group, *Figuring Out Education Spending*; Nan Robertson, 'Subtracting Anxiety from Math', © 1977 New York Times Company, reprinted by permission; Christina Robb, 'A Fable Made Up of Numbers', © *Boston Globe*; Michael Robin, 'Black Babies: A Right to the Tree of Life', *The Nation* magazine/The Nation Company, Inc., © 1984; Joel Rogers, 'The Politics of Voter Registration', © *The Nation*; Dexter Tiranti, 'A Pill for Every Ill', © *New Internationalist* November 1986; 'Truth in AIDS Tallying', *Boston Globe*; 'What's in a Number?', *In These Times*; Julian Weissglass, 'Small Groups'.

Note to Readers

This maths text is very different from a traditional maths text. And, at first, this way of learning maths may seem harder. There's a lot of reading and writing; there are a lot of choices you have to make; there are ambiguous problems; you are asked to create and evaluate problems. You may want to resist; you may want to relearn maths the same way you were taught it in school. However, as I argue in the first part of this text, a key reason for people's 'anxiety', anger, and avoidance of maths is a set of misconceptions they learned in school. That's why it's important to relearn maths in a different way.

My students' lives are filled with full-time work (or with full-time searching for employment), with families, and with full-time school. Obviously, much resistance to the hard work required to relearn maths comes from the realities of their daily lives. But, in addition, I feel some resistance comes from students having internalized the dominant society's views that 'the intellectual activity of those without power is always characterized as non-intellectual' (Freire and Macedo, 1987, p. 122). My students often don't realize that they already know much about maths – the 'academic maths' decimal point, for example, is the same as the point used to write amounts of money (for example, pounds and pence, or dollars and cents). And my students are often reluctant to realize that, in spite of many gaps in their knowledge of maths, they *already* think logically about maths. As educator Paulo Freire pointed out: 'Our task is not to teach students to think – they can already think; but to exchange our ways of thinking with each other and look together for better ways of approaching [learning about the world]' (Freire, 1985a).

Two main goals of this text are to help you gain the self-confidence to realize that you can relearn maths and, simultaneously, to help you realize that *right now* you already do intellectual work in maths. A main focus of this text, then, is to help you understand, as a participant in a cultural-action literacy programme in Africa remarked: 'Before, we did not know what we knew. Now, we know that we did know and that we can know more' (Freire and Macedo, p. 167). To accomplish this, you will look at error patterns to analyse the correct, as well as incorrect, aspects of the reasoning behind their solutions. You will look at how other people's reasoning has been misinterpreted as being illogical. And you will be urged to take your intellectual work in maths seriously. Intellectual work is hard work – relearning maths needs to be approached with discipline and persistence. With that attitude you will relearn maths in such a way that you will appreciate its underlying structures, and will be able to use it as one way of understanding or, as Freire would say, reading the world.

I end this note by asking you to reflect on the meaning of the following excerpt from a text used in São Tomé and Príncipe after these African countries won their freedom from the Portuguese:

> The Act of Studying: I
> It had rained all night. There were enormous pools of water in the lowest parts of the land. In certain places, the earth was so soaked that it had turned into mud. At times, one's feet slid on it. At times, rather than sliding, one's feet became stuck in the mud up to the ankles. It was difficult to walk. Pedro and Antonio were transporting baskets full of cocoa beans in a truck to the place where they were to be dried. At a certain point the truck could not cross a mudhole in front of them. They stopped. They got out of the truck. They looked at the mudhole; it was a problem for them. They crossed two metres of mud, protected by their high-legged boots. They felt the thickness of the mud. They thought about it. They discussed how to resolve the problem. Then, with the help of some rocks and dry tree branches, they established the minimal consistency in the dirt for the wheels of the truck to pass over it without getting stuck.
>
> Pedro and Antonio studied. They tried to understand the problem they had to resolve and, immediately, they found an answer. One does not study only in school.
>
> Pedro and Antonio studied while they worked. To study is to assume a serious and curious attitude in the face of a problem.
>
> The Act of Studying: II
> This curious and serious attitude in the search to understand things and facts characterizes the act of studying. It doesn't matter that study is done at the time and in the place of our work, as in the case of Pedro and Antonio, which we just saw. It doesn't matter that study is done in another place and another time, like the study that we did in the Culture Circle. Study always demands a serious and curious attitude in the search

to understand the things and facts we observe.

A text to be read is a text to be studied. A text to be studied is a text to be interpreted. We cannot interpret a text if we read it without paying attention, without curiosity; if we stop reading at the first difficulty. What would have become of the crop of cocoa beans on that farm if Pedro and Antonio had stopped carrying on the work because of a mudhole?

If a text is difficult, you insist on understanding it. You work with it as Antonio and Pedro did in relation to the problem of the mudhole.

To study demands discipline. To study is not easy, because to study is to create and re-create and not to repeat what others say.

To study is a revolutionary duty! (Freire and Macedo, 1987, pp. 76–7)

Dedication

This book is dedicated to all the people who have helped form my political consciousness, in particular:

My mother, Jean Frankenstein, who is neither a leftist nor an activist, but who first explained to me the insanity of racism and the logic of racism under capitalism, when at ten years of age I overheard a neighbour refer to my family by saying, 'Leave it to the kikes to bring the niggers around'

My father, Eli Frankenstein, whose ability to make friends with everyone, from people he meets on the check-out line in the supermarket to people he meets riding on the bus, inspired my outgoing involvement with and love of people, without which political insight is reduced to empty rhetoric

My uncle, Matthew Chaves (1914–83), with whom I loved to discuss politics, whose radical political vision encouraged me to develop a critical analysis of the conditions shaping our society, and whose wonderful sense of humour gave me the perspective I needed to keep struggling for that vision

Notes on Contributors

Marilyn Frankenstein has been teaching maths to adults since 1978. For the previous decade she taught maths to junior high school, high school and young adult students in New York City and New Jersey. She currently heads the Department of Applied Language and Mathematics, works at the College of Public and Community Service, University of Massachusetts, Boston. Her previous publications include *Basic Algebra* (Prentice-Hall, 1979) and numerous articles on maths teaching. She is a member of the *Radical Teacher* editorial collective and *Science for the People* editorial advisory board.

Henry A. Giroux is Professor of Education at Miami University, Ohio. His latest books include *Teachers as Intellectuals* (Bergin & Garvey, 1988) and *Schooling and the Struggle for Public Life: Critical Pedagogy in the Modern Age* (University of Minnesota Press, 1988).

Europe Singh has taught maths since 1972 in Inner London Educational Authority (ILEA) schools. During 1983–6 he was seconded to ILEA's anti-racist strategies team. Co-ordinator of the Campaign for Anti-Racist Mathematics, he has taught at South West London College since 1987.

Ubiratan D'Ambrosio is Professor of Mathematics (and since 1982 Pro-Rector for University Development) at the Universidad Estadual de Campinas, Brazil. Having taught at several universities in the USA, he is currently President of the Latin American Society for the History of Sciences and Technology.

Introduction

There are many levels on which we have to fight to make our individual and collective lives meaningful. This book argues that knowledge of basic numerical and statistical concepts is important in these struggles.

In a recent personal struggle, this truth was dramatically reinforced for me. My ability to ask the most basic mathematical questions was a key factor in decisions I had to make concerning completely conflicting recommendations from 'top-of-the-field' specialist doctors on how to treat my newly diagnosed, rare cancer. Some of the significant questions I asked were, 'How many people were involved in the study on which you are basing your recommendation?' 'What is the average age at the time of diagnosis of this cancer?' The answers to these questions and other analyses of available statistical data (Gould, 1985) – along with eventually finding a few supportive, caring doctors – helped me to feel confident that I made the correct choice in this literally life-or-death situation.

In our collective struggles, we are continually making life-or-death choices. Unfortunately, the media, the 'general knowledge climate' of assumptions about 'the way the world is', and widespread 'number numbness' often obscure this (Hofstadter, 1982). Even more often they obscure the information available upon which we can make those choices. A particularly chilling example of this occurred during the panel discussion following the showing of *The Day After*, the television movie about nuclear holocaust. Dr Carl Sagan rebutted Secretary of State Shultz's contention that the USA was reducing its nuclear arsenal by pointing out that official US government figures showed a planned increase in the number of strategic nuclear warheads, from 9,000 to 14,000. The moderator, Ted Koppel, quickly interjected, 'Carl, forgive me. Let us leave this in general terms because I must confess statistics leave my mind reeling, and, I suspect, everybody else's too' (Manoff, 1983).[1]

One of the objectives of this text is to help you understand mathematics in a way that will enable you to use that knowledge to cut through the 'taken-for-granted' assumptions about how our society is structured, and to act from more informed choices about those structures and processes. Ted Koppel only needed to understand the maths concept of 'comparing numbers' (see Chapter 12) in order to realize that Carl Sagan's argument is based on demonstrating that under Reagan the number of nuclear warheads has *increased*. Since many adults in his TV audience already understand that 14,000 is greater than 9,000, viewers not scared off by Koppel's using the word 'statistics' might have wondered why he stopped the introduction of numerical evidence backing up Sagan's argument.

Unemployment and welfare are two other examples of issues illuminated by the kind of basic numerical analysis you will feel confident to undertake after completing this text. Basic statistics dramatically illustrate that, contrary to stereotypes about 'laziness' and 'scrounging', people do want jobs. Unemployment and wage reductions have made people so desperate as to compete intensively for even the most menial jobs. For instance, in 1975 in Watts (a low-income, largely black neighbourhood of Los Angeles), a hospital advertised for forty low-wage attendants, and 6,000 people applied (*Newsweek*, 18 August 1975); and in 1980, when the Social Security administration of Baltimore, Maryland, advertised seventy-five entry-level positions, 26,000 people picked up application forms (*New York Times*, 21 September 1980).

Statistics also show that, contrary to much popular belief, welfare in our society is overwhelmingly distributed to the rich. For instance, in the mid-1970s the maximum AFDC (Aid to Families with Dependent Children) welfare payment to a four-person family was $5,000, while the average tax loophole for the richest 160,000 taxpayers was $45,000 (Babson and Brigham, 1978, p. 37). According to a 1984 report by the Congressional Budget Office on the combined impact of Reagan tax benefit cuts since 1981, households with incomes of under $10,000 lost $390 that year, and families with incomes of over $80,000 gained $8,300. Furthermore, in 1981, $510 million of the USA's collective tax money was spent on new airports for

private pilots so they would not land their planes at large commercial airports (*In These Times*, 4 March 1981).

RELEARNING MATHS

The kind of mathematical literacy needed to clarify issues, to understand the structure of society, and to support or refute opinions is more than the ability to calculate. It is the ability to understand what numbers mean – how the number system is constructed to describe aspects of the world. It is to understand what operations (addition, subtraction, multiplication, division) do to numbers – how performing a numerical operation can change data in ways that clarify or obscure reality. It is also the ability to know what kinds of basic statistical questions to ask to deepen understanding of particular issues, and the ability to present data to change people's perceptions of those issues. Volumes 1 and 2 of this text provide a thorough development of the basic maths needed to become mathematically literate.

Volume 1, Part One focuses on overcoming your 'number numbness' or maths 'anxiety' in several ways: by analysing the misconceptions about learning that block your understanding of math; by reframing your fear into anger at a system which has led to the mathematical disempowerment of large numbers of adults; and by suggesting learning activities that will deepen your growing understanding of mathematics.

Volume 1, Part Two focuses on the meaning of numbers. You will practice reading, writing, comparing, and rounding whole numbers, fractions, decimals, and percentages *before* solving any problems using addition, subtraction, multiplication or division. In this way you learn to approach problems more carefully. As most of these problems do not require doing an operation, you overcome the 'panic' approach of reading fast, picking out the numbers and performing some operation just to 'do something'.[2] Some of the problems do involve basic operations, and others imply operations; for example, you might only be asked which number is larger, but you could also subtract to find out how much larger. So, you will get a preview experience of doing operations before systematically studying their meaning in Volume 2.

Volume 1 ends by illustrating how groups use maths in critical, oppositional ways. By providing a list of such groups and their publications, it encourages students to investigate additional examples and to develop their own ways of using maths for progressive social change.

Volume 2 treats problem-solving and problem-creating (asking questions) by using real-life data and recreational applications. You will learn the abstract meaning of each operation, so it will no longer be a mystery to guess what to do with the numbers in a problem; you will be confident of which operations to perform to clarify specific numerical information. The conclusion of Volume 2 explores the critical education theories of Paulo Freire, Henry Giroux and others in the context of the mathematics learned in both volumes of this text. This material is included in order to demystify learning processes, and to emphasize the empowering purposes of learning, so that you can build on your experience learning mathematics to learn anything you need in the struggle for personal and collective liberation.

OTHER LEARNING OBJECTIVES

As well as teaching this kind of mathematical literacy, this text also aims:

- To help you realize that you already know some maths (for example, if you can deal with money, you understand a lot about decimal arithmetic).
- To make you confident that you are capable of learning any mathematics you need to know in order to become mathematically literate.
- To help you learn about the learning process, so that you can control your own learning through such actions as self-evaluation and formulation of specific questions to clear up misunderstandings.
- To counter the myth that 'slow' means 'stupid'.

This text helps you to practise the slow, careful thinking needed to examine critically the structure of our society. As Ira Shor points out, 'the sensory flood from media joins the rush of daily life to make minds work too fast to do close reading of texts and critical scrutiny of ideas' (1978, p. 122). Because small visual changes in the symbols can totally change the meaning of a mathematical expression, you are forced to slow down your perceptions. Because many of the application problems in this text either contain more information than needed, or require that you find additional information, you get practice in examining and searching for data, rather than immediately spitting out an answer. And, because this text involves you in creating your own maths problems, you get practice in examining many different possibilities in order to determine which questions you can ask and answer. For this reason it aims:

- To illustrate the recreational, 'fun' aspects of mathematics (for example, magic that works because of maths concepts, art based on maths patterns).
- To increase your background knowledge about a variety of important social, political, economic, and cultural issues that do not get well covered in the mass media. The data used in this text are taken from real life and presented to encourage you to ask more questions about the 'facts' behind many of our society's assumptions.
- To help you realize that no teacher is an 'expert' with all the answers. Because the applications are drawn from a wide variety of areas, it is very likely you will raise subject-matter questions that your teacher cannot answer. The Appendix suggests how you can become skilled at searching for information to answer your own questions – how you can become 'critical co-investigators in dialogue with the teacher' (Freire, 1970a, p. 68).
- To motivate you to fight for progressive change and to know when and how basic statistics can be an important part of that struggle.

OBJECTIONS TO THIS KIND OF MATHS LEARNING

A number of pedagogical and philosophical objections have been raised concerning my approach in this text. The major ones concern:

- The large amount of reading and writing in this text: won't this confuse diagnosing whether students are having trouble with the maths or with the reading and writing?
- The 'non-mathematical' character of many of the problems: if you're not focused on exercises such as 'adding fractions by finding the LCD', are you really learning mathematics?
- The 'non-neutrality' of the entire presentation: by including only one point of view, isn't this text 'propaganda'?

To answer the first two concerns, I argue that if you are having trouble with understanding this text, you are having trouble with the reading *and* the writing *and* the mathematics. Trying to separate those ways of knowing the world leads, at best, to a mechanical facility with maths rules. Once a student asked me in her journal, 'Why can't you just give us the *a*, *b*, or *c* and all you have to do is add, subtract, multiply, or divide?' I answered that a basic mathematical understanding of the meaning of the operations requires that the numbers themselves have concrete meanings. What is learned by simply being able to move the numbers around correctly, unless one knows which moves to make in any given situation and what the results of those moves mean? Reading and writing and mathematics 'are aspects of understanding, and theories that attempt to account for them outside their interactions with each other run the risk of building reductive models of human understanding'.[3]

To answer the third concern, I argue that no knowledge or education is ever neutral. Most schooling and daily life bombard us with messages supporting the status quo. Even trivial maths applications, like finding the total from a grocery bill, carry the non-neutral hidden message that it's natural to distribute food according to individual payment. Even traditional maths courses which provide no real-life data carry the non-neutral hidden message that learning maths is separate from helping people understand and control the world. Texts such as this present another point of view. I believe the best we as teachers and writers can do is to tell you our own biases and encourage you to use numbers to examine the opposite views to those you may hold. As William Profreidt points out, just the opposite happens in most educational settings where there exists

> . . . a silly neutralism in which teachers believe they are just presenting facts and avoiding opinions or value statements. They avoid value statements because . . . they do not believe that such statements are susceptible to rational inquiry and verification. Of course, in practice they are transmitting a set of values, but one which is not identified as such, and hence is not open to critical inquiry. (1980, p. 477)

PROBLEMS

1 Read 'Truth in AIDS Tallying' (*Boston Globe*, 8 September 1987). Discuss the main point and what kinds of mathematical knowledge you need to understand the editorial.

Truth in AIDS tallying

The official definition of AIDS has been broadened by the Centers for Disease Control, the federal agency in charge of monitoring the epidemic. As a consequence, the tally of AIDS cases kept by the CDC will rise substantially over the next few months.

But the rise still will not reflect the total number of AIDS cases in the United States since the epidemic was recognized in 1981. Cases have been underreported from the outset. The CDC's count of AIDS cases is low by 40 percent, Daniel Haney, the Associated Press's science writer in Boston, reports.

The new definition will correct only 15 percent of that miscount. One in four AIDS cases will still go unreported. But the public needs to understand that the definition does not mean the disease has changed or that there really are more cases.

Although CDC's announcement of the new definition notes the limitation and lack of usefulness of the old one, more should be learned of why CDC stuck to the outmoded definition so long. Some critics are convinced the reason was less scientific than political, pandering to the Reagan administration's opposition to any step that called attention to AIDS.

When the epidemic began, a strict definition of the disease made sense, to ensure that only authentic cases were reported. This is important in the outbreak of any disease, and enables a careful tracking.

Originally, an AIDS diagnosis required that the patient had a strange skin cancer, Kaposi's sarcoma, or a rare pneumonia, *Pneumocystis carinii*, and that these disorders developed after the person's immune system for unknown reasons had broken down and was no longer protective against disease.

After the AIDS virus was identified, the definition required that a test show the patient had been infected. And a few other cancers and bizarre infections were added to the syndrome, which still defined only acute and terminal AIDS.

Meanwhile, thousands of Americans fell prey to an early, chronic form of AIDS – swollen glands, night sweats, fever, lethargy, weight loss and recurrent bouts of illness – that was called AIDS Related Complex (ARC). Even though the victims also had impaired immune systems and were carrying the AIDS virus – and many died – they were excluded from the AIDS count.

This was especially hard on AIDS-afflicted children who, though they were deathly ill, had symptoms that did not fit the strict CDC definition. As a result, they were denied coverage for costly AIDS tests and treatments under Medicaid. To get around this problem, pediatric AIDS was redefined a year ago. The new AIDS definition will include anyone who is showing signs of one or more of 18 infections and 11 cancers, along with evidence of the AIDS virus. This will encompass nearly all ARC victims as they move into active stages of the disease.

The definition also accepts three forms of AIDS that were not previously counted, though they have been well known: AIDS-related tuberculosis, AIDS-related dementia, and AIDS-related emaciation, or wasting syndrome. The last is called 'slim disease' in Africa, where it is a common type of AIDS.

Although the new definition will quickly raise the AIDS count to nearly 50,000 cases and 25,000 deaths, the count will still be short by about 25 percent. It is not unusual for reportable diseases to be underreported, but a strong argument can be made for seeking greater accuracy about AIDS.

An analysis of funds needed to limit the spread of AIDS, issued in August by the General Accounting Office, cited the uncertainty of government estimates on AIDS cases – and costs. The US Public Health Service predicts 270,000 cases of AIDS by 1991 at a direct medical care cost of $8.5 billion. But a study by Rand Corporation, a California think tank, calls that a low-range estimate and posits a high range of 750,000 cases by 1991, at a cost of $112 billion.

In the face of a lethal epidemic that keeps running ahead of efforts to combat it, Congress and state officials need accurate AIDS information quickly. Impediments to AIDS reporting, whether bureaucratic or political, are downright dangerous.

2 What follows below is, first, an example I included in an article I submitted to a mathematics journal, and second, the changes its editor made, without telling me, after accepting my article. After reading both versions of the problem, discuss these questions:

- What is the difference between how I wrote the example and how the editor rewrote it?
- Why do you think they felt that my phrasing was not neutral?
- Why do you think they felt that their phrasing was neutral?
- Why is their phrasing not netural? (And, why was I glad the current editor agreed with me and used my version of the problem?)

My version

A chart from the Census Bureau (*Statistical Abstracts of the United States 1982–83*, Chart No. 709, p. 439), which can apply various operations with whole numbers, decimals, fractions, or percents.

Household income (current dollars)

	1967	1976	1977	1978	1980	1981
White families	7,449	13,289	14,272	15,660	18,689	20,153
Black families	4,325	7,902	8,422	9,411	10,764	11,309

1 Clarify the information in this chart by computing some relevant comparisons. (The student can discuss whether, for example, subtracting to find the income gap, or finding the percent of white income that blacks are paid, better clarifies the data.)

2(a) Write a brief statement of your opinion about the causes of this income difference.

2(b) List the kinds of numerical data that you feel would support your opinion. (The goal of this exercise is to make students aware of how people use and find numbers to support their arguments. For example, to support arguments that, in spite of progress, discrimination against blacks is still a significant problem, students can use unemployment rates and infant mortality rates. According to the *Guardian* newsweekly, 20 October 1982, official black unemployment in September 1982 was 20.2 percent compared to official white unemployment of 10.1 percent. According to the *Statistical Abstracts of the United States 1979*, the non-white infant mortality rate was 21.7 per 1,000 births in 1977, compared with white infant mortality rates of 12.3 per 1,000 births in the same year. Students can also investigate why there is no separate category for black infant mortality.)

Edited version

A chart from the Census Bureau can be used to apply various operations with whole numbers, decimals, fractions, or percents.

Household income (current dollars)

1967	1976	1977	1978	1980	1981	
White families	7,449	13,289	14,272	15,660	18,689	20,153
Black families	4,325	7,902	8,422	9,411	10,764	11,309

Students are asked to:

1 Clarify the information in this chart by computing some relevant comparisons. (For example, students can discuss which better clarifies the data: subtracting to find the income gap or finding the percent of white income that blacks are paid.)

2(a) Write a brief statement of your opinion about the causes of this income difference.

2(b) List the kinds of numerical data that you feel would support your opinion. (The goal of this exercise is to make students aware of how people find and use numbers to support their arguments. For example, to support arguments that in spite of progress discrimination against blacks is still a significant problem, students can use unemployment rates and infant mortality rates. Alternatively, they may compare black and white employment rates when educational level is held constant.)

3 As the previous example shows, seemingly small changes in language can make significant changes in meaning. In the edited version, the use of the word 'alternatively' implies that looking at employment rates, holding education constant, would show that racial discrimination is no longer a problem. In fact, this implication is completely false. The Council on Interracial Books for Children (p. 7) reports statistics that show black college graduates have a greater jobless rate than white high-school drop-outs! And, according to the *Statistical Abstracts of the United States 1986* (Chart 753), white elementary-school graduates have a greater median income than black high-school graduates.[4]

The terms we use for such comparisons also involve mathematical assumptions. In my own use of language, I am careful not to use the word 'minority', because it denies the reality that people of colour are by far the majority of the world's population. According to the *Statistical Abstracts of the United States 1986* (Chart 1463), Africa and Asia alone comprise 70 percent of the world's population.

In this textbook I ask you to learn to analyse what numbers are used for what purposes, but I also want to encourage you to be thoughtful about the words you use to describe the world. Here is an interesting example of language differences: progressive people in the USA use the general term 'people of colour' to describe Afro-American, Asian, Native American (etc.) peoples, whereas in Britain they use the term 'black'. Which term do you think works best, and why?

4 Read the following excerpt from an interview that Les Levidow (my editor at Free Association Books) conducted with Grazyna Baran

about her experience of teaching maths to adults in London. Write briefly about which of her students' reactions you relate to, and why; which reactions you disagree with, and why.

Grazyna Baran (GB): Before starting the job I was told that the students felt terribly nervous about their maths ability. If I hadn't been told that, I might have taken a different approach to the statistics course.

After thinking about this, I decided it would be a good idea to help the students revisit their earlier maths experience. With that view, I started off the whole course with a diagnostic test. I made it absolutely clear to them what I was doing and why I was doing it. I explained that I was mainly interested in their reactions to the test. Apart from helping them find out where they stood arithmetic-wise, the test would also evoke all the feelings that maths usually brings out in people.

I naïvely hoped that, by giving them the test and explaining my purposes, they might actually relax, be less nervous and just do it. But in fact their reaction was almost automatic. As soon as I handed out the test paper, even before the students looked at it, they all reacted negatively. Some of them went blank, and their eyes went cold. The women said things like, 'Oh, I can't do this, I'm so stupid.' Everybody's nervous reaction was different, but everybody's was noticeable.

Les Levidow (LL): What was that all about?

GB: The test questions started off with a straightforward addition, then a straightforward subtraction, a slightly more complex subtraction, then went on to multiplication, division, then a minus times a minus – all basic arithmetic tasks. It was remarkable how people did the tasks at the level at which they'd last done them at school. There were some people who had not gone beyond the junior school level, where people are taught long division in a very mysterious way; they're told to follow certain instructions without understanding why. If they have a remainder, then they are told to write it down as a remainder rather than as a fraction. This is because many junior school teachers traditionally haven't had the confidence themselves to teach fractions; instead they just reproduce their own knowledge of arithmetic the way they were taught it. And they think that junior school students can't cope with fractions. So if someone writes down a remainder, it means they've never got beyond their experiences and impressions of arithmetic at junior school level. They may have experienced fractions in other contexts, but they remain stuck at the earlier level when they do a test.

There were other interesting features of the diagnostic test. Many people couldn't handle a minus times a minus; signs became totally confusing to them. Many people hadn't learned their times tables; they were counting on their fingers or just giving up because it was so difficult for them. Again, I sensed the damage that might have been done around maths. It's almost as if they said, 'Well, if I can't do my times tables, then I'm stupid, and therefore I can't do maths.' They were stuck because of that.

LL: So many didn't even go through with the test?

GB: A lot didn't get further than about ten computations, out of the thirty or forty on the test.

LL: And not just because of working slowly – a lot were paralysed?

GB: Yes. When we had our discussion about it, many of those things came out. A common feeling was, 'I was told that if I couldn't do my maths, then I might as well forget it, because getting your maths and your English are the most important things.' There was one man who had been considered the brightest in his junior school. But when he got to secondary school, he found he couldn't cope with the maths. It was such a source of persecution for him that he didn't bother with anything else in school, and he became a truant.

LL: The warning became a self-fulfilling prophecy?

GB: Yes, and there are many of these cases. Another man, considered especially bright, came under great pressure from his father to succeed in maths and ended up with a similar failure. It's interesting that it was two men with the same kind of story. When we did some exercises with graphs and bar charts, he sat there unable to put pen to paper. When I asked him why he wasn't doing it, he said, 'Well, I know how to do it. I can see what the issue is, and I'm quite sure I could do it, but there's something in me preventing me from putting pen to paper. All I can see is that bastard, my father, standing over me.' I could see the anger welling up in him: his fists and teeth were clenched, his lips pursed. So people were coming up against memories that were very painful.

I think the men were suffering more than the women, because at least the women were able to express their frustration and fears. Having done so, they then got on with the test and said: 'Oh, it wasn't so bad after all. I can do it.'

LL: You mean, when you handed out the test, the women's reaction was more overt?

GB: Yes.

LL: Do you think they'd been pushed as much as the men when they were younger, but dealt with it differently?

GB: I think that the feeling that 'Everything is lost unless . . . ' was put much more on to boys than girls. The women in the class were unconfident and dithery, but once we confronted those feelings, they were more able to revisit them, improve, and develop some confidence. While the feelings seemed more deeply ingrained in the men, who remained paralysed.

We had an interesting discussion about maths. One student said, 'I don't bother working it out. I just look at the numbers and get a feel more or less what the answer is.' Because I was interested in validating people's experiences, I said, 'That's pretty good. A lot of mathematicians actually do that. In fact, it's more important to have a feel for numbers than to get everything absolutely right. You can use calculators for the arithmetic.' But when I looked at what he'd done, I was appalled because he was utterly lost.

LL: You mean it wasn't even wild guessing?

GB: On the test it was. But afterwards, when we did bar charts and so on, he was lost. In the discussion we talked about why anyone would want to collect statistics in the first place, and for what purposes. A lot of the discussion was sceptical about what information statistics would really give you, if any. So there was a lot of awareness on that front. But this person wasn't at all willing to engage with statistics at some level.

It was surprising that he actually attended the course – nearly every session. Yet he saw statistics so much as a conspiracy against working-class people that he didn't want to know about it at all. So he was dealing with a contradiction. By attending, he showed that he did want to know about it, but in the course he closed off his mind to it. That told me that there was a lot of pain there, which he wasn't able to revisit.

LL: 'Conspiracy' suggests a conscious political view.

GB: That's what he had. He's very much a Marxist but wasn't at all clear about statistics. I think there are times when it's important to gather some information and make sense of it; statistics are not necessarily an exercise to be used against people. I argued that it could be empowering to understand how people are diddled by statistics and how we could use them in better ways.

I feel that my argument didn't make much headway; he blanked himself off because there was another level that he found very threatening. I think he was rationalizing all that with a political framework, shielding himself from a past that he didn't want to confront. I should add that he was black. And that's important: from my experience of secondary schools, wherever there's streaming [US: tracking], the lower stream is almost entirely black. So I think there's a real connection between implicit and explicit forms of racism and this man's feelings about maths. Of course it's also a class issue. There's the feeling that 'Either I can do it or I can't. If I'm stupid then it's not worth bothering to try.'

LL: From your anecdotes, it seems that maths had a special, powerful significance, in that the parents portrayed it as the make-or-break subject. Perhaps there's something about maths in this society that presents itself in those terms: that either you get it or you don't. Is it different from other subjects in that way?
GB: That's a difficult issue. I think there are parallels among different subjects; many people feel just as alienated from other sciences as from maths. Maths is special because it acts as a discriminator in a much wider way. Everybody is expected to do it, and there are so many jobs that require maths as an entry qualification, sometimes to a higher level than is justified by the work.

I'll tell you one interesting strategy that I used. I told the students that a lot of the actual calculating was a mechanical task. Some of them said, 'I don't know my times tables, I've got to count on my fingers, and by the time I've counted I've probably missed and got it wrong.'

So I told them that there's no reason why they can't use a calculator: they're cheap and easily available. But it was almost as if they didn't want to use calculators because the arithmetic was a challenge and they were beating themselves over the head with it. It was as if using calculators was cheating; they felt, 'If you have to use one, then you're stupid.'

To encourage them to use calculators, I set a large part of the follow-up work of the first diagnostic test to go through a few very simple rules – brackets, notation, and so on. These were simple things that perhaps they never noticed before. The purpose was to see some connection among mathematical operations; for example, the connection between decimals and fractions, and that a minus times another minus number gives you a plus.

But some of them didn't want to accept that the skill in using a calculator was in knowing how to use it – the rules of number that tell you what buttons to press. I explained that calculators save you from having to do the tedious bits. In these days you don't really need to know your times tables to do calculations quickly. Of course it helps, because it gives you some idea of number relationships. But if that worries you, then you can use a calculator, and it doesn't mean you're stupid.

In fact I said that, if as a child you found learning your times tables really tedious and boring, and you couldn't be bothered, that's a sign of being pretty clever really – because what a bloody boring thing to do. If you can find a way of surviving without having to do that, that's a pretty clever thing. So this is how I tried to empower them – by reframing their sense of suffering from not having learned those tables.

In that reframing, I said that the calculator is a 'stupid machine'. You can use it because number-crunching is a pretty repetitive, automatic task – which is why it can be computerized. You're not a machine: you needn't do the boring bits.

But there was much resistance to this idea. One student said, 'Oh, if I can't do what a stupid calculator can do, then I really must be stupid.' So even after my attempt, they were still using the idea of the calculator to bash themselves. Whichever way I tried it, the reaction was, 'Oh, if it's that, then I'm even more stupid.' So I threw up my hands in exasperation and asked why they wanted to believe that: were they trying to prove that they were stupid?

LL: So their reaction attests to the power of internalized feelings – so strong that the best logical argument still can't overcome them.

Part One

Mathematics: anxiety, anger, accomplishment

1 Mathematics anxiety: misconceptions about learning mathematics

Learning how to drive at the age of twenty-seven was a significant educational experience for me as a teacher. I learned at first hand that motivation, self-confidence, and persistence can overcome fear. For me, the motivation was beginning an excellent job which required driving. I needed a great deal of self-confidence in order to ask questions that most people would think of as 'stupid'. (When my instructor asked me to start the car during my first lesson, it was only self-confidence that got me through the embarrassing moment of admitting that I had no idea how to accomplish that task.) I needed persistence not to give up after my first set of lessons ended, as I still could hardly drive, and I couldn't even dream of ever being able to parallel park closer than two feet from the kerb.

Because the title of my unwritten autobiography is *Fear of Driving*, I have always had strong empathy with other people's fears. This feeling, and my eighteen years' teaching experience, have convinced me that most maths 'anxiety' is overcome when one's 'maths autobiography' is understood, when the myths that schooling and society have created about learning are dispelled, and when new ways of learning are applied to mathematics.

Most people have had negative experiences with mathematics. As Peter Hilton, a prominent US mathematician, has argued, 'the experiences which most children have of mathematics through their mathematics lessons and through the mathematics texts which they use are . . . very likely to lead them to seek to avoid mathematics' (1980, p. 176). He describes the features of current maths curricula which he feels are most responsible for maths avoidance:

- *Rote calculations*, which do not increase understanding and are unnecessary because of the availability of hand calculators (whose use requires knowledge of all the important computations such as estimation).

- *Memory dependence*: 'There is absolutely no point in learning the proof of Pythagoras's theorem if one does not understand the proof; and if one understands the proof, there is no point in learning it' (p. 183).
- *Unmotivated exercises*, such as many fraction addition problems 'where there is no indication of any situation in which such a skill would be necessary or even desirable' (p. 183).
- *Spurious applications*: In real life one wouldn't use subtraction to solve the problem, 'Juan goes shopping, His mother gives him $1.50 and he spends $0.89. How much does he have left?'; rather, one would add what is left in Juan's pocket (p. 184).
- *Authoritarianism in mathematics education*, where 'students are told what to do, told that it will turn out to be useful, told that it works, told that they will understand it later' (p. 185).
- *Tests which 'assume mathematics can be divided into tiny water-tight compartments*, so that one can test individual items . . . and mathematics is just a discrete set of masteries of such items' (p. 186).

I feel that these aspects of typical maths curricula lead most people to develop misconceptions about learning. This section focuses on examining those misconceptions and suggests ways of overcoming maths 'anxiety' by 'learning about learning'. Although the following discussion of common misconceptions pays attention to affective factors, it concentrates on the cognitive blocks to math learning:

• ***Misconception***: 'I'm the only one who didn't learn elementary school maths when I should have.' Many people are under the misconception that there are specific times when certain things should be learned, and that, therefore, they are somehow inadequate because they can't do mathematics that they 'should have learned in elementary school'. They also think their feelings are unique, that 'everyone else understands and feels comfortable using maths'. Laughing about my driving fears and sharing others' learning fears should certainly help relieve anyone's feeling of inadequacy about learning a skill 'later in life'! Any current gaps you have in your maths knowledge do not reflect any lack of intelligence or humanity; they just mean you have some new skills to learn. And this text will help you learn those skills in an 'adult' context. The basic maths is presented and practised in connection with individual and collective interests, from jogging or music to unemployment statistics.

• ***Misconception***: 'I'll never be able to learn maths.' Not only do many people mistakenly feel embarrassed about not having learned basic mathematics in elementary school, but they are also under the misconception that no matter how much they study and practise, they will never be able to do mathematics. Sheila Tobias, who popularized the term 'maths anxiety' in a 1975 *Ms* magazine article, emphasizes that 'what is particularly pernicious about the math mythology is the notion that from the outset people either have or do not have a "mathematical mind"' (p. 8). I felt the same way about driving: before I learned how to drive, I could never even mentally picture myself behind the wheel, controlling the car. If I learned how to drive, you can learn maths!

• ***Misconception***: 'Smart people do maths fast, in their heads, and in one sitting.' Many people are under the misconception that they are stupid if they have to think too long about a mathematics problem. On the contrary, solving maths problems involves slow, careful thinking and 'pacing'. This text suggests many techniques for working on problems. But whatever methods you choose, it is important to understand that each of us has a point at which we get stale or frustrated. When we can no longer think clearly about a problem, we should temporarily stop working on it. This is not 'giving up'; we can come back to that problem later with fresh insight and either solve it or formulate some clear questions we need answered before we can solve it. Sometimes, instead of resting, work on other problems leads to an insight which helps us solve this initial obstacle. It is a misconception to think that each topic must be completely understood before beginning to study the next topic. Learning does not occur along a linear path; learning occurs in layers, where it is necessary to tolerate ambiguity until later knowledge clarifies and deepens our understanding of previously studied topics.[1]

• ***Misconception***: 'I'm stupid if I make a mistake or ask a question in class.' Many people are under the misconception that 'smart' people never make mistakes and do not have to ask questions: 'smart' people figure out everything correctly, on their own. On the contrary, people learn from mistakes. No attempted solution is *all* wrong; analysing what aspects of the solution are correct, and pinpointing exactly what confusion led to the mistake, will reinforce what you already know and will help you not to repeat the error. Also, asking questions is a very important part of learning. This text emphasizes the development of question-creating skills as a key step in your controlling your own learning. There is always some aspect you do understand, even if it's only the meaning of some of the words in the problem statement. When you just say to a teacher, 'I don't understand,' rather than asking a specific question, you give control to the teacher to determine what you need to learn. If you feel embarrassed asking questions, just think of me when – during my first driving lesson, having no idea how to start the car, and being told to 'pump the gas and turn the key' – I asked my driving teacher where the gas-pumper was!

• ***Misconception***: 'There is only one correct answer to each maths problem.' Many people are under the misconception that since '2 + 2 = 4', real-life maths application problems also have only one correct answer. On the contrary, in real life the assumptions made, and the accuracy with which the data are collected, both affect the answer. As Munir Fasheh argues:

> 'One equals one' is a mathematical fact, but its description and interpretation and application differ from one situation to another and from one culture to another . . . one dollar in 1970 is not equal to one dollar in 1980 . . . antagonistic feelings and different opinions emerge when we say for example that 'women are equal to men'. Strictly speaking, then, 'one equals one' does not have true instances or applications in the real world. (1982, p. 5)

Unemployment statistics provide a clear example of how differences in interpretation result in different 'correct' answers. For instance, official British employment statistics for 1985–6 (Londoners' Living Standards, Report 2) list the numbers of unemployed people in Britain at 4.75 million. However, if we were to include all people in forced retirement who want jobs, the national unemployed total would be 6.5 million. We could also include people who are no longer actively seeking employment because there aren't jobs available locally. And there are many other categories whose count would give other 'correct' answers to the question, 'How many people are unemployed in Britain?'

• ***Misconception***: 'There is only one correct way of solving each maths problem.' Discussing how unemployment statistics might be collected (for example, from employers, from unemployment offices, from a population sample of workers) can introduce the idea that there are many methods for solving a problem. Sometimes one method is more reasonable, although this can often depend upon one's perspective.

For example, is it a 'correct' method to rely on data taken from the registration records at unemployment offices, or will many unemployed, but non-registered workers, be overlooked? Governments think this is a correct method, in spite of the claim that 'many workers fail to register with the Employment Services Agency, either because they are not entitled to unemployment benefit or because they do not wish to claim it . . . This most obviously affects married women who opt not to pay insurance contributions . . . [Also] the problems of non-registration . . . are particularly serious amongst immigrant groups' (Hyman and Price, 1979, pp. 228–9).

Even in basic computation, there are many choices of methods. For example, there are many ways to add:

Method A (standard paper-and-pencil column addition):

$$\begin{array}{r} {}^{1}54 \\ +\ \underline{39} \\ 93 \end{array}$$

Method B (mental addition, thinking of 39 as 40 − 1):

$$\begin{aligned} \text{You think, } 54 + 39 &= 54 + 40 - 1 \\ &= 94 - 1 \\ &= 93 \end{aligned}$$

Method C (mental addition, treating the tens' place and units' place separately):

$$\begin{aligned} \text{You think, } 54 + 39 &= 50 + 4 + 30 + 9 \\ &= \underbrace{50 + 30}_{} + \underbrace{4 + 9}_{} \quad \text{(because changing the order in addition doesn't change the answer)} \\ &= 80 + 13 \\ &= 93 \end{aligned}$$

Analysing the advantages and disadvantages of each method will help you to evaluate which one works best for you.

• ***Misconception***: 'The teacher is the only one who can tell me the answers.' Many people are under the misconception that checking one's own answers from a key is cheating. On the contrary, how can you learn much by working on problems one day and not finding out whether or not you were correct until days later? If you made an error early in your work, since you didn't evaluate as you proceeded, that error will have been repeatedly reinforced in later problems. Whereas, if you checked your answers from a key as you finished each problem, the correct ones are reinforced; the incorrect ones can be analysed, and you can either figure out how to solve the problem correctly, or you can formulate a question about what is confusing.

PROBLEMS

1(a) Write briefly about your earliest maths memory. (Don't worry about grammar and spelling – focus more on remembering details.)

(b) Discuss which of Hilton's criticisms of misconceptions relate to your earliest maths memory.

2(a) Complete the following sentence with as many words as occur to you: *When I think about learning maths, I feel . . .*

(b) Ask at least four people of various ages and educational backgrounds to complete that sentence.

(c) Discuss the answers and the reasons behind them with each person.

3(a) Which analysis of the misconceptions discussed in this section makes the most sense to you? Write briefly about why and how it relates to your experience.

(b) With which analysis of the misconceptions discussed in this section do you disagree? Write briefly about why and how it differs, from your experience.

4(a) My difficulties in learning how to drive provide just one example of the fact that teachers are not 'experts' who can easily learn everything. The main learning difference between teachers and students is that teachers know how to go about learning in general. This book attempts to empower you by teaching you general learning techniques as well as mathematics. At this time, I suggest you focus on how you have learned in the past:

(i) List various skills you have such as the ability to swim or cook, or spell, which were initially difficult to learn.

(ii) List the techniques you used to learn each skill.

5 It is important for you to realize that you already understand some things about mathematics. For example, the decimal point in '$5.42' is the same decimal point in decimal arithmetic.

(a) Brainstorm a list of activities in which you use maths.

(b) Next to each item in the list, write the maths skills that you need to understand in order to be able to do the maths required for that activity.

6 One way to illustrate why solving mathematics problems involves slow, careful thinking and why a mathematics book cannot be read like a novel is to investigate how small visual changes in the symbols can totally change the meaning of a mathematical expression.

(a) What is the visual difference between these two arithmetic expressions?

$$3 \times (4 + 6) \qquad \text{and} \qquad (3 \times 4) + 6$$

(b) Using the rule that operations in parentheses are performed first, how does the small visual change affect the meaning in the above two expressions?

7 The following are samples from time-and-motion studies, conducted by General Electric, and written up in a 1960 handbook to provide office managers with standards by which clerical labour should be organized (Braverman, pp. 320–1):

Open and close	*Minutes*
Open side drawer of standard desk	0.014
Open centre drawer	0.026
Close side drawer	0.015
Close centre drawer	0.027
Chair activity	
Get up from chair	0.039
Sit down in chair	0.033
Turn in swivel chair	0.009

When I asked my students for their reaction to these time-and-motion studies, most of them felt they had to solve a maths problem. (See Notes: Introduction, n. 2.) So they became confused and frustrated. This is a good example of slowing down when you read anything with numbers. In order to comprehend the point of these studies, you only need to know one thing about the numbers. What is it?

8 One of the central focuses of this text is helping you learn how to ask specific questions about maths problems you feel you can't solve. This is very difficult; your problem-creating skills will be gradually developed throughout the entire text. Right now you may feel what Plato expressed in his dialogue *Meno*:

'A person cannot enquire about that which he knows or about that which he does not know; for if he knows, he has no need to enquire; and if not, he cannot; for he does not know the very subject about which he is to enquire'. (Stempien and Borasi, 1985, p. 17)

(a) Rephrase, in your own words (and in non-sexist language), the dilemma that Plato posed.

(b) Try to imagine some ways out of this dilemma. Then, write a list of suggestions for how to formulate questions about new or confusing topics.

9 Read the following article by Nan Robertson in the *New York Times* (1 May 1977).

'Subtracting anxiety from math

'I thought I had a terminal case of math anxiety. For as long as I could remember, I suffered from panic, embarrassment, resentment and self-loathing when faced with the simplest problem of mathematics. Quick! What is two-thirds of 40? How do you figure the 15 percent tip on a lunch tab of $12.35? How do you compute miles per gallon on your Volkswagen?

'"Dummy!" I would scream at myself in the void where my brain was supposed to be. "You can't do it!" And then I would pull down the little window shade in my mind until the problem went away, or I begged someone else to solve it for me.

'Then, this spring, I found Mind Over Math.

'A friend of many years, for whom figures hold no terrors ("Eileen," I would whine, in a scene repeated many times over, "what is 5 percent of 1,000?") sent me in a roundabout fashion to Stanley Kogelman and Joe Warren, who run Mind Over Math in New York. They are two young mathematicians who started these group therapy workshops last June. Stan has a master's degree in clinical social work, both are college teachers with doctorates in mathematics and both have been through psychotherapy. Both are very, very comforting.

'In the past year, they have tried to "demystify" math for lawyers, film producers, carpenters, housewives, high school dropouts, bank officers, store managers, even Harvard Business School students. The age range has been from 25 to 60. Most have had a college education. Two-thirds have been women. The Kogelman–Warren observation – and it is backed by studies – has been that math anxiety disproportionately afflicts females, who also tend to avoid math more. Society has long encouraged this, with the belittlement, "There, there, little girl, don't bother your pretty little head about numbers." Well, after years of taking this guff and, by now, angry at myself, I dialed 662–6161 and enrolled myself in Mind Over Math. The cost was $75 for five Wednesday night sessions of 80 minutes each.

'We met, my group and I – all seven of us jittery – in a cozy room at the Berkshire Hotel on Madison Avenue at 52nd Street, with cookies on a corner table and a coffee pot perking away.

'That first session, we talked about ourselves, digging memories about family and classroom traumas and trying to pinpoint when we began thinking we were dummies. Jim remembered a math teacher with a stop-watch who said "Go!" when the tests began. "They never told us to spell faster," he said. The Kogelman–Warren technique, familiar to anyone acquainted with group therapy, is to draw people out gently and encourage them to share their problems and fears with other members of the group. As almost always happens, members of the group discover that few human experiences are unique.

'"I didn't know other people felt the way I did," Paulette said when the session was over. "I was very, very embarrassed and very secretive about coming here, but I'm not now."

'All of us sailed out of that first 80 minutes in a state of unfocused euphoria – maybe there was hope for us after all.

'A week later, at the second session, Stan and Joe passed out pieces of paper with words and numbers printed on them and instantly I felt that old, familiar panic.

'The problem began: "If two men can dig a ditch . . . What a dumb problem! I'm sure I have more important things to worry about, especially today. It's really been cold but today it finally warmed up to around 32 degrees. Just a few days ago it was only 5 degrees, way below freezing, so now 32 degrees feels warm . . . except in my apartment."

'One look at the paper had instantly triggered me. I felt overwhelmed, upset, confused, chaotic, I wanted to flee, the way I always had.

'But when I calmed down and read the first problem over, I realized the answer was perfectly simple. I subtracted 5 from 32 to get 27 – the number of degrees that the temperature had warmed up. It was so easy I thought I couldn't have identified the problem correctly.

'"Why was I so anxious?" I wondered aloud. "This isn't a test. Nobody's going to judge me. I've been out of school for more than 25 years and I felt like a frustrated, guilty, muddled kid."

'On each of the succeeding three sessions, I found myself more comfortable – able to read the problems, break them down and get the answers. "We aren't going to teach you how to *solve* problems," was the refrain throughout, "but we are going to teach you how to *approach* solving problems."

'My approach became: "Whoa, Nan. Slow down and isolate the little rascal the way you do with most other things in life."

'Stan and Joe taught us something else – that there is no "right" way to solve a math problem so long as you arrive at the solution.

'The funny thing was, I felt no triumph at getting the right answers. None of us did. And therein lay one of the secrets of Mind Over Math.

"You see?" said Stan. "Math – it's no big deal."'

(a) Identify the main idea and supporting details.
(b) About how much money do Stan and Joe make from these sessions?

(Solve this problem using at least two different correct methods, and get at least three different correct answers, stating which assumptions you used to arrive at each answer. What missing information would help determine the actual amount they make?)

Note: You may not be able to solve this problem yet. If that's the case, do not panic. Instead, do as much as you can. For example, locate the numerical facts that you need in the article. Try to pinpoint the exact spot at which you get stuck. Try to formulate a specific question whose answer will enable you to solve more of the problem. For example, 'I could figure out the problem if each session were 1 hour, but I don't know what to do with the 80 minutes.'

10 The misconception that 'there is only one correct answer to each math problem' was recently illustrated by one of my students, Wendell Houston. I asked the class to write 'seven and eight hundred five thousandths' in mixed number/fraction form. Wendell wrote

$$7\ \frac{800}{5{,}000}$$

(a) What other correct answer is there to this problem?
(b) How might we distinguish between these two correct answers in the English written names for these two mixed numbers?

2 Mathematics anger: mathematics is not useless and boring

People are justifiably angry when required to learn mathematics in school because maths has been traditionally taught in ways that make it seem useless and boring. This text counters those beliefs by illustrating the usefulness of mathematics in your roles as citizen, consumer, and worker; and by illustrating (although more briefly) the fun of mathematics through its interrelationships with more 'recreational' topics, from puzzles and magic to art and literature. Furthermore, it explores some of the reasons why so many people feel maths 'anxious' and argues that it is not accidental that so many people are mathematically 'disempowered' by our school system.

PROBLEMS

Read through all these problems in order to get some idea about ways math is used in the world and about aspects of math that may be interesting to you. Then work on the problems that interest you.

1. You are probably much less mathematically disempowered than you have been led to believe by previous negative experiences in math classes. Look through a daily newspaper or weekly news magazine. Record instances of mathematics that interest you, and try to explain how the numbers are used to support and/or clarify the main point of each article. If you are unsure about the meaning of the numbers in a particular article, try to pinpoint exactly what confuses you. Ask someone to help you clear up what you don't understand.

2. Many citizens' groups use numbers to argue for social change. For example, Boston's Coalition for Basic Human Needs presents statistical analyses of living costs in Massachusetts to argue for adequate cost-of-living increases for AFDC (Aid for Families with Dependent Children) recipients. For another example, London's

Counter-Information Services (CIS) a group of journalists, trade unionists, and statisticians, reconceptualizes information in official corporate reports, at the request of workers at the companies involved. In one case, CIS (1972) used Rio Tinto Zinc Corporation's data that 42 percent of its profits were made in South Africa, whereas only 7.7 percent of its assets were located there, along with additional information CIS researched, to support the charge that these high profits came directly from the low wages paid to black miners.

List a citizens' group organized to fight for change in your neighbourhood and describe how that group uses or might use numbers in its work.

3 Read the following excerpt from 'The politics of voter registration' (Rogers, 1984).

'But can voter registration really make a difference in electoral politics? The numbers say it can. Consider past Presidential elections. A shift of 7,000 votes in 1960 would have given Richard Nixon the Presidency eight years sooner than he got it; a shift of 8,000 would have denied Jimmy Carter the office in 1976. And even in the case of Ronald Reagan's "landslide" in 1980, the numbers are revealing. Despite an overwhelming electoral victory (489 to 49), Reagan's percentage of the popular vote was only 50.7, just 0.6 percent greater than Carter's in 1976, and exactly 10 percent less than Nixon's in 1972. In many states, Reagan's margin was very thin (see the tallies for Arkansas, Delaware, Kentucky, Maine, Massachusetts, Mississippi, South Carolina and Tennessee . . .), so small mobilizations of "reliably Democratic" voters – blacks, for example, who vote Democratic at approximately a 90 percent rate – would have made a big difference. This was especially evident in the South. In Alabama, Reagan's margin was 17,642, while the number of unregistered blacks was 272,390. In Arkansas the comparable figures were, respectively, 5,123 and 85,383; in Mississippi, 11,808 and 130,910; in North Carolina, 39,383 and 505,711; in South Carolina, 11,456 and 292,000; in Tennessee, 4,710 and 157,673. Elsewhere, the story was similar. In New York, Reagan's plurality was 165,459; the number of unregistered blacks was 893,773. Assuming a 9-to-1 Democratic split on the black vote, if 20 percent of the unregistered black population had signed up and voted [New York's 41 electoral votes] would have been delivered from the hands of the Reaganauts.'

(a) Briefly state the main point of the excerpt.
(b) How do the numbers clarify and support that main point?

4 Read the following excerpts from Oxfam's response to popular myths about food and population. Briefly state the main point of

each response and discuss how the numbers clarify and support the arguments. Then write a brief statement expressing your opinion about world hunger; list the kinds of numerical data (not the actual numbers) that would clarify and support your opinion.

Myth 1

There is. The world is producing enough food to feed every man, woman and child. Enough grain is produced to provide every person with more than 3,000 calories a day.

And that's without counting beans, fruit and vegetables. But that's not all! Enough food is available in those very countries where so many people are forced to go hungry.

In Mexico, where 80 percent of children in rural areas are undernourished, livestock consume more grain than the entire population – livestock that is then exported to the USA to be made into hamburgers.

In Bangladesh, after the 1974 floods, 4 million tons of rice remained in storage because people could not afford to buy it.

Overpopulation – *the* bogey word. But which countries are overpopulated? Those which consume the most, like the USA with 6 percent of the world's population consuming over 25 percent of the world's resources? Or one of the poor countries?

Those which are most densely populated? Holland has 1,079 people per square mile, India 613, Britain 565 and Brazil 39. And what of those countries where thousands of people die every year because they haven't food to eat, but where there is enough cultivated land per person? In Africa, south of the Sahara – an area where famine is commonplace – there are 2½ cultivated acres per person. That's more than in the USA and 6 times more than in China.

SO... It isn't just a question of numbers. The hungry countries are not the ones which are the most densely populated, or which consume the most. But they *are* the countries where the poor aren't given access to land to grow food, or money to buy it.

5 List situations in which you use maths in your role as a consumer.
 (a) For each situation, describe the aspects of the maths that you understand.
 (b) For each situation, describe any maths you would like to know that would help you make better decisions in those situations (for example, knowledge of what interest rates mean in loaning money).

6(a) List some tasks at your work in which you use math, and describe the kinds of maths involved.
 (b) List some tasks at your work in which you don't use maths, and try to come up with some ways in which you could use maths in those situations.

7 In addition to focusing on the usefulness of maths, both volumes of this book challenge your creativity through recreational maths puzzles and patterns. They attempt to spark your interest in some of the cultural aspects of maths through maths-art, including maths-magic tricks, maths-visual-art, maths-music and maths-poetry. 'Culture' and/or intellectual recreations are not something reserved for those with advanced graduate degrees, or for those from the upper classes. 'Culture' is what all of us create from nature to help us lead productive and enjoyable lives. Of course, the focus is real-life applications, so there are only a limited number of recreational

problems, and the 'maths culture' here is very subjective: it represents the play uses of maths that I love. I hope my examples will encourage you to discover and create other maths-related contributions to our culture.

(a) List your hobbies and other 'non-work' activities.

(b) List any maths that is involved in each of the above activities, and describe the specific uses of maths in some detail.

8 Various computations form incredible number patterns. As you learn more maths, we will analyse why some of these patterns occur. For now, in each of the following groups of computations, discover the pattern and write three more lines in the pattern.

(a)	(b)	(c)
$3 \times 37 = 111$	$999 \times 2 = 1998$	$9 \times 9 = 81$
$6 \times 37 = 222$	$999 \times 3 = 2997$	$99 \times 99 = 9801$
$9 \times 37 = 333$	$999 \times 4 = 3996$	$999 \times 999 = 998001$
$12 \times 37 = 444$	$999 \times 5 = 4995$	$9999 \times 9999 = 99980001$

9 Many magic tricks can be explained by maths. For example, pick your favourite digit and multiply it by 3. Then multiply your answer by 37.

(a) What magic surprise do you get?

(b) To see why this trick works, do these multiplications in a different order. That is, first multiply 3×37. Now try to explain why you get a magic surprise when you multiply that result by your favourite digit.

10(a) Describe the illustration on the frontispiece of this book. Can you figure out how the numbers and words are related?

(b) Why do you think I chose that art for the frontispiece of this book?

(c) Why did I subtitle the book *A Different Third R – Radical Math(s)*?

11 The illustration below was created by Karen Shaw, a contemporary artist who often works with numbers. Her work converts words to numbers and vice versa to give us a new way of viewing our number-saturated world. She explains her use of numbers as follows: 'I designate a numerical equivalent to each letter of the alphabet according to its position (A = 1, B = 2, C = 3 . . . Y = 25, Z = 26) . . . A word is "spelled" out numerically and added to reach the sum of the word . . . Numbers are transcribed into words of the equivalent sum and collected in a numerically ordered vocabulary. I call this process Summantics. Since a particular number can equal the sum of various words, the choice is determined by rhythm, sound, imagination, grammatical structure or my mood at the moment . . . I seek the poetry in register receipts, the messages of passion, fear and aspiration that are encoded on the lineup on the line of

scrimmage.' Karen Shaw's Summantics dictionary contains over 10,000 words listed in numerical order. Explain how the words and numbers are related in this illustration.

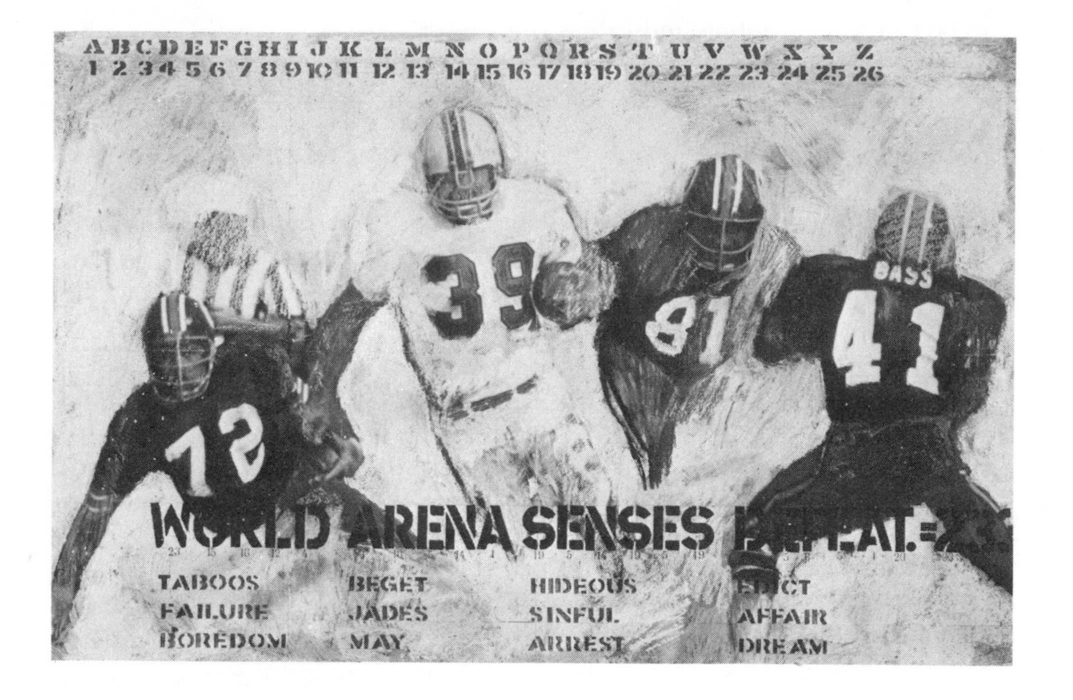

12(a) Describe the illustration on the title page of Part One. What aspects of it seem mathematical to you?

(b) Why do you think the artist, Maurits Escher, entitled it 'Liberation'?

(c) Why do you think I chose it for the title page of Part One?

13 I have always felt uncomfortable calling people 'maths *anxious*'. The effect of such a label is contradictory. On the one hand, students are initially relieved that their feelings about mathematics are so common that educators 'have a name for them'. On the other hand, the label can focus the problem inward, 'blaming the victims' and encouraging solutions directed solely *at* them. The label can direct attention away from the broader social context of how these misconceptions about learning come to be so universally believed, about what purpose might be served by such widespread 'maths anxiety', especially among women.

Volume 2 discusses these issues in more detail. It suggests that exploring the underlying causes of maths anxiety can be an effective way for you to turn your inward 'anxiety' into an outward constructive anger that will motivate you to seek maths accomplishment. What do you think about these ideas?

3 How to use this book

When I first learned how to drive, there seemed to be a thousand things I had to remember at once, from checking the traffic light and turning on my directional signal, to turning on my directional signal correctly and looking both ways before making a left turn. Very gradually, these things became automatic. I no longer have to remember consciously to take my foot off the gas (the accelerator) and put it on the brake when I see a red light – I now do that automatically. As you are learning to use mathematics, it may seem to you that there are a thousand things to remember at once – from the meanings of the place-values through to which keys to press when on your calculator. Just as driving became automatic to me, using maths will become automatic to you.

This book uses an immediate feedback/self-evaluation approach to help you to learn how to use numbers to analyse and fight to change the conditions of your lives and of society. Starting with Part Two, each chapter consists of: brief text explanations; solved problems (drill, thought, application, and sometimes recreation), with similar problems for you to try; and, similar unsolved problems followed by a review quiz. Answers to unsolved problems are at the end of each chapter. You are not expected to be able to do the solved problems *before* you read them – they are a way of presenting the material to you gradually, each solved problem introducing a new concept, building on what you learned in the previous problems. After you study a solved problem, do the corresponding 'You Try These' problem. Either your correct method will be positively reinforced, or your particular mistake can be analysed and you can note specific questions to ask to clarify your misunderstanding. Also, you can use the format of this book to learn how to evaluate your own understanding. For example, if you are familiar with the meaning of whole numbers, take the review quiz before reading through that chapter. Check the answers at the end of the chapter; you may decide to review only the application problems, or you may decide that you understand the topic well enough to go on to the next chapter.

The most difficult thing to do when you want to learn something is to know exactly what you already understand and what you still need to find out. Once you can evaluate your own knowledge and can find information and people to work with, you can learn anything you want.

PROBLEMS

1(a) Flip through the text. What catches your eye? Why?
(b) How is this text similar, how is this text different, from other maths texts you have used in other learning situations?

2(a) Read the Contents table and list the topics with which you feel some familiarity.
(b) List the topics you think will be easiest, and those you think will be hardest, to learn.

3 List the key features of this text. Briefly describe why each feature is included.

4 The problems in this text are divided into five types: drill, thought, application, recreation, and project.
(a) Drill problems provide practice in manipulating numbers in isolation from concrete applications. You need to be able to perform the various operations of addition, subtraction, multiplication, and division in order to get answers to the maths applications.
(b) Thought problems pose questions to encourage you to spend time reflecting on the structure of our number system. Maths is not a mysterious, mystical language which only an élite can understand. The thought problems are intended to deepen your understanding and thereby demystify the interrelationships among numbers.
(c) Application problems introduce you to the variety of areas which an understanding of maths can clarify. They present real-life data usually omitted or made inaccessible in the schools and in the mass media. These data, and further investigations you undertake, will deepen your critical analysis of the structure of our society. Whenever possible, the maths questions are not just exercises, but are intended to clarify, support, or refute the given information. These problems also introduce you to a variety of problem-solving strategies.
(d) Recreational problems introduce you to some fascinating math topics – patterns, puzzles, magic and art. Usually students tracked (streamed) into basic or vocational maths classes spend all their time drilling on the minimum skills needed to pass state 'competency'

exams. This inhibits thinking and encourages an apathetic, negative attitude towards learning. Of course, since the text focuses on real-life applications, there are only a few recreational problems, but they serve as an important introduction to creative thinking and the pure enjoyment of learning.

(e) Projects are suggestions for topics to investigate in more detail. Some are intended to give you experience in using maths to document a critical analysis of the conditions of everyday life, and others are intended to challenge you with more in-depth recreational, creative thinking. They are clearly optional; you may want to just read them until you finish studying all the material in the text. Then you will I hope create your own projects to use maths to study the specific issues that relate to your activities.

Problem 4 is to find an example of each type of problem in this text.

5(a) List two things you like, or think you will like, about this text.

(b) List two things you dislike, or two questions you have, about this text.

4 Why using a calculator helps

For most adults studying mathematics in college or in community learning groups, time is a privilege.[1] Given that limited time, I recommend you focus on the more creative aspects of mathematics, from understanding the structure of the number system to solving and posing complex maths problems. Calculation is the least important maths technique – rules are involved whether you use paper and pencil or a calculator. You can choose which rules to learn; I recommend the calculator ones, because they can usually be learned more quickly.

Some people feel using a calculator is 'cheating'. This is a misconception. The creative challenge in solving problems is to understand what you are being asked to find; to figure out which numerical information is needed for the solution and how to find it; to determine which operations (addition, subtraction, multiplication or division) to perform; and, to be able to estimate the answer so you can check if your answer makes sense. The actual computations can be done 'long hand' by pencil and paper, or in a much shorter time by calculator.

The calculator that I suggest you purchase (before beginning Part Two, although it will be most useful in Volume 2) should cost under \$10 or £6, contain a one-year guarantee, look something like the following diagram, and contain the features listed below. Extra features, such as trigonometric functions, are unnecessary, hard to use, and expensive.

1 The display should show up to eight digits. It should be a liquid crystal display (LCD). These LCD calculators use small disc-shaped batteries which last much longer than other batteries.

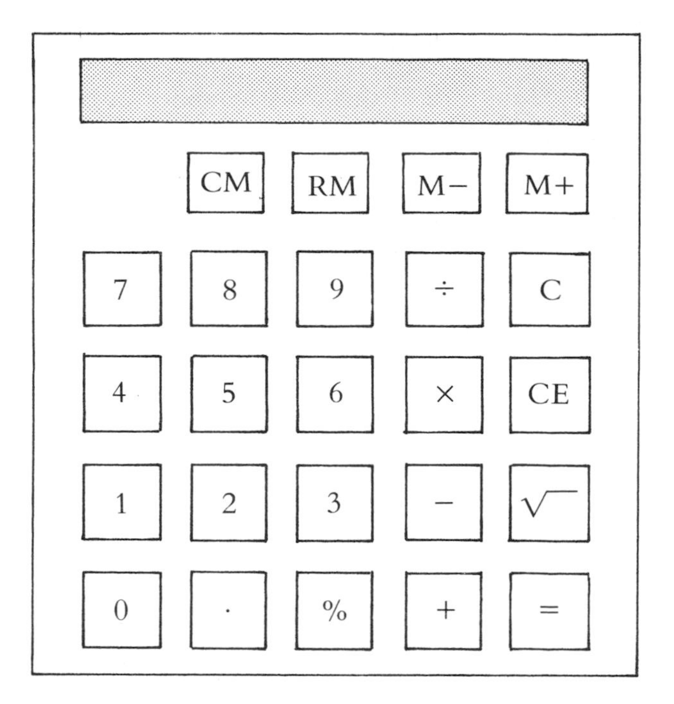

2 [C] clears the display and sets it at 0.

3 [CE] clears just the last number you entered on the display.

4 [·] is the decimal point.

5 [0], [1], [2], [3], [4], [5], [6], [7], [8], [9] are the ten digits from which (with the decimal point) we can write any number in our decimal number system.

6 [+], [−], [×], [÷] are the four arithmetic operations of addition, subtraction, multiplication, and division.

7 [=] is pressed to get the answer to your calculation.

8 [%] is used in percent problems. This key is optional and functions differently on different calculators.

9 [√] is used to find the square root of a number.

10 The four memory keys allow your calculator to remember one number while you are computing with other numbers. They are particularly important in working with fractions.

[CM] clears the memory.

[RM] is used to recall the memory. When you press it, the number stored in the memory appears on the display.

[M+] adds the number you have just calculated to the number stored in the memory.

[M−] subtracts the number you have just calculated from the number stored in the memory.

PROBLEMS

1 What is the main idea of this cartoon about our most sophisticated calculating machine?

"Does 63592187653.4215400532 look right to you?"

2(a) Briefly state the positive and negative feelings you have about using calculators.

(b) Interview a few other people to find out how they feel about calculators, and write a brief summary of your discussions.

3 Read Isaac Asimov's 'The feeling of power', in Clifton Fadiman's *Mathematical Fantasia* (1958, pp. 2–14).

(a) What is the main point?

(b) Do you agree or disagree and why?

5 Keeping a journal

Keeping a maths journal, or diary, is an effective way to overcome maths 'anxiety', a method for clarifying which learning techniques work best for you, and an excellent aid to understanding mathematics concepts and problems.

Reflections from a variety of my students' journals illustrate how writing about the problems you're having learning maths can relieve tension and help you discover how you learn best:

- I can see now how important that is [to get everyone to participate in the class discussions . . . for everyone to help each other] . . . The fact that another student can answer the question that one presents is an encouragement . . . to put forth a little more effort because if another student can understand enough to answer, then you can understand.

- I am beginning to see why it is so important to do your assignments after a discussion in class. It is very important to me as a helpful tool to follow up on the discussions because what I feel that I have comprehended in class will only be reinforced with the homework. So, if there is any misunderstanding or confusion, I can bring it up in the next class and not feel left out or as if something has gone by me and I will never be able to do it.

- I must spend more time listening than trying to write notes as you talk – I am so afraid I will miss something you are saying – but I found today that *listening* is more important – many things are coming back to me.

- The one important thing I am learning about algebra is that if I get stuck on a problem I shouldn't just quit as I have in the past but stay calm and try to think about what we have gone over in the classroom, the way each step was figured out and *try* to remember the rules . . .

- I think I know part of the reason math is difficult for me. I seem to do math in a disorganized manner . . . While doing my homework I became conscious of the amount of scrap paper I used. The figures were all over the paper and at times I couldn't figure out what part of the problem I was working on and what I had figured out before . . . I'm going to have to find a method of keeping things organized.
- The concept, about the teacher [as authoritarian] who has power because he/she possesses the answers, while the student doesn't, made a big impression on me. I feel that much of the retreating, which I did in school, was because I felt most teachers were very judgemental . . . The concept of self-evaluation, is a very freeing one.

The journal may even help you probe the deeper causes of your maths 'anxiety', as with one of my students who traced his maths fears:

> Growing up, I was raised by my grandparents who had definite roles (i.e. man breadwinner – woman housewife), so I was conditioned to a lot of sexist stereotypes. Math fits into that mold. I feel that my fears of math, and thinking that I could never do it, have always been a subtle reflection on my masculinity. Not that I ever doubted being a man, but maybe a little less of a man when I would stop reading an article or report when the author started throwing statistics or other numbers around.

The journal can also help you discover solutions to problems through the process of thinking about them in writing. For the purposes of restating the problem and of recording the work (rather than keeping everything 'in the head'), writing is an important aspect of problem-solving. By rewriting your problem in your own words,[1] by summarizing the given information, by writing down related questions, by performing operations and evaluating whether the results provide new information useful to the solution, you can learn more about how to solve the problem.

This learning process has been described by Arthur Powell, a colleague who teaches maths and maths education at Rutgers University, and José López, a student who worked with him in a participatory research project on using journal writing to learn mathematics. They conclude:

> Writing, because the writer and others can see it, allows one to explore relationships, make meanings, and manipulate thoughts; to extend, expand or drop ideas; and to review, comment upon and monitor reflections . . . The more learners are involved in choosing language, the more they are engaged in constructing and reconstructing meaning, making sense of mathematics for themselves. (Powell and Lopez, 1989)

Therefore, even if you don't initially *know* how to solve a problem, the process of writing can help you decide what directions to explore.

Also writing, as an extension of memory, can alleviate a form of maths anxiety that my colleague Lucy Horwitz (1978) has identified as 'cognitive discomfort':

> [This anxiety] comes, not in anticipation of having to solve a math problem but in the course of doing so, after an initial struggle . . . In solving a mathematical problem, we must keep a certain amount of information in mind and we must process that information. The act of rehearsal can help us keep the information in mind, but it will make it impossible to simultaneously process it. If we process, we cannot rehearse and lose our grasp of the essential facts. We feel like jugglers, trying to keep too many balls in the air, and it is the effort of this task that causes the cognitive discomfort which, if prolonged, can be felt as intense activity. (pp. 1–3)

Writing down all the numbers involved in a problem with notes about their meaning and interrelationships is one way to reduce the load on short-term memory and, therefore, to reduce cognitive discomfort.[2]

PROBLEMS

1 As you work through this book and/or attend class, spend about five minutes, three or four times per week, writing about past experiences learning maths, how you felt in class, how you attempt to understand this book and do the problems, how you use maths outside school, how you feel about your progress, and any other related areas.

2 Various topics for diary entries are found in the problems of previous chapters asking you: to write about your earliest maths memory; to discuss the aspects of the chapter you found useful and why; and to comment on articles about mathematics learning, and so on. Sharyn Lowenstein, a colleague who has studied journal-writing, has suggested a number of other interesting specific assignments:

(a) Describe one moment in the maths class which was significant to you. It may have been when you understood what had been confusing. Or it may have been a moment of embarrassment. Try to use very specific detail.

(b) Draw an abstract representation of your maths class and then comment about this picture.

(c) Choose one of your maths teachers who gave you a hard time years ago. Create a dialogue between you and that teacher. Tell the teacher what you now need to say. Create responses for this teacher.

3 Another reason for writing about mathematics learning is given by Margaret Stempien and Raffaella Borasi (1985):

'Writing involves an organization of our ideas and notions about a topic. This may prove to be a difficult task, but with a worthwhile pay-off: the realization of relationships and connections existing among the bits and pieces that we have separately learned.' (p. 16)

Each time you learn a rule or maths concept, rewrite it in your own words in your journal (diary).

4 Yet another reason for keeping a journal is to give you a concrete record of your maths progress. Often people denigrate their successes and focus on what they cannot yet do; keeping a maths journal will help you to realize that you can do now what one month ago you thought was impossible. As one of my students commented:

'One thread runs through this journal, the same one that has run throughout most of my life. I cannot accept or recognize my own intelligence and ability to do things capably or to solve problems. Looking at my record at school, I see that I *have* done well . . . but I keep thinking that I must have fooled them!'

So it makes sense to look back over your journal after some time has elapsed in order to see the progress you've made and to learn from the feelings and experiences you have recorded. Every few weeks reread your journal entries and list a few things you learned over that period.

5 Read the following excerpt from a journal written by a student in a colleague's math class (Buerk, 1981). Comment about what you think she is saying; what you agree with and why; and what you disagree with and why.

'And on the eighth day, God created mathematics. He took stainless steel, and he rolled it out thin, and he made it into a fence forty cubits high, and infinite cubits long. And on this fence, in fair capitals, he did print rules, theorems, axioms and pointed reminders. "Invert and multiply." "The square on the hypotenuse is three decibels louder than one-hand clapping." "Always do what's in the parentheses first." And when he finished, he said, "On one side of this fence will reside those who are good at math. And on the other will remain those who are bad at math, and woe unto them, for they shall weep and gnash their teeth."

'Math does make me think of a stainless-steel wall – hard, cold, smooth, offering no handhold, all it does is glint back at me. Edge up to it, put your nose against it, it doesn't take your shape, it doesn't have any smell, all it does is make your nose cold. I like the shine of it – it does look smart, intelligent, in an icy way. But I resent its cold impenetrability, its supercilious glare.

'Anger, frustration, resentment, panic – I'm trying to write about what math is, and what I come up with is how it makes me feel. I resent it right now that I can't think of an intelligent definition of math – math is what you do with numbers when you're little, and what you can't do with letters when you're older – if you're me. Arithmetic I know – that deals with countable things, in one way or another, and the operations make sense to me. The idea of a fraction or a percent makes sense to me, too, but multiplying and dividing fractions is a slippery business. What I learned in math each year became progressively less useful and more senseless. I went as far as algebra in ninth grade. (Geometry is not math, contrary to popular opinion – I like geometry.) Algebra was the end of math for me – I have utterly no way of imagining what math is like beyond algebra. I'd be hard put to say exactly what algebra is – letters instead of numbers, but to what end I don't know. I saw some progression in it, because the formulas kept getting more letters in them until finally we got to the quadratic equation. The formulas got longer, and I presume new, according to some grand design, but as I learned them, they seemed discrete – no interlocking, no building, just more things to memorize, so that with luck I could fill in the numbers and get an answer. You had to be lucky twice for every algebra test problem – first, you were lucky if you remembered the formula, and second, you were lucky if you added or multiplied without making a "careless" mistake. Being prone to bad luck seemed to be an unchangeable quality, like being left-handed.

'What strikes me as being particularly strange and yet even now completely typical of my encounters with math was my bland expectation of not understanding algebra. What I did expect was to "see" it sometimes and other times not. I would look at a problem and notice how the numbers went into their slots. Sometimes they looked OK. Sometimes they didn't look OK, even when they were right, but that was math. It was the only subject that just seemed to have certain intrinsically impenetrable aspects. I grew accustomed to hearing a sort of math-babble, words that attached themselves to nothing, numbers that seemed arbitrarily plunked into equations. I'm sure I copied it all down, too – to memorize, in case I'd need it on a test, or just on the chance that because of my very dutifulness, I would be granted sudden enlightenment. I obviously never asked questions.

'So what is math? I can't get very far with an answer. I have heard math spoken of as a language. I can't imagine it. Language is personal. Language carries thought forward, discovers thought, creates thought. Whatever it is that I know as math doesn't do those things.' (pp. 159–60)

This is an excellent journal — you're keeping good track of your feelings/experiences

Class #6

Would you like to read this to the class? I understand if not, but I think this discussion would be good for the whole class.

→ I know that I ended my last entry into this journal as saying that, "I am ready to tackle the next class," but I wasn't. I was very tired and became bored at the very start of the class. I have to learn to control my feelings of being critical of other people problems in Algebra. I found myself thinking of the questions that some of the others asked as being elementary. I just assumed that if I understand, everyone should.

it's good that you catch yourself — it's hard to be patient with others' problems the key to remember is that you want others to be patient with your problems To make it more interesting for yourself, try to explain the concepts to others — it's interesting to see exactly what step different people don't understand and to try ways of getting through.

Some of the problems I did have a little difficulty doing them but I did not mention it in class because I felt that I would sound stupid or should I say unable to comprehend what was being said.

at this point you should realize that this is not so!

Finally I began to fight the feelings that I had about other people problems and ~~[illegible]~~ started being more ~~attentive~~ of what was being asked. I began to understand more and more and at one point, the question that I wanted to ask was answered. So, it wasn't so stupid after all.

Great! you not only learned math, but learned about the learning process!

6 Try to find a partner with whom to share your journal. I comment on my students' journals, offering encouragement, alternative solutions or perspectives, and explanations of how their remarks on learning maths often apply to learning in general. Students' comments are very helpful for my lesson planning since they show: what students understand and what confuses them; which mathematics skills and concepts are most helpful to them; which mathematics topics are most interesting to them; which teaching techniques work best for them; and, how they proceed as they work on their mathematics problems. Work out a situation where you and your journal-partner learn from each other's writings. The page of handwriting printed here is an example of a student's journal with my comments.

7 Another way to use writing to help you learn maths was suggested to me at a conference at Roxbury Community College. Some maths teachers there create a collective set of class notes. The students, on a rotating basis, are responsible for taking class notes and reproducing their notes for the entire class. In this way everyone can learn from a discussion of the notes – what is more clearly stated in the notes than, say, in the text and why; what is missing that was learned in the class and how that could be added, and so on. If you can't arrange this in your class or learning group, try to find one person with whom to exchange notes about your learning.

6 Working with others

Most traditional school courses are taught using what Paulo Freire calls 'banking' methods: 'expert' teachers depositing knowledge in the blank minds of their students; students accepting without question that memorizing the required rules will get them future dividends (1970a, chapter 2). Often these methods make people believe that they have no knowledge to share with others and that they are cheating if they do their school work with others. *Schooling in Capitalist America* (Bowles and Gintis) presents detailed research to document that these methods also play an important role in preparing people to compete for boring jobs over which they have no control, and training people to follow rules obediently, without understanding, and to take their 'proper place' in society, without questioning.

By working with others as you learn maths, you can begin to break the patterns of social relationships which traditional schools often reproduce. Try to work with a group to teach each other maths: share solutions to problems, evaluate which of the many correct methods is clearest to you, and analyse and learn from each other's mistakes. When you teach someone in the group, first try to determine what s/he knows. Often, because someone else is using a different approach than the one you understand, her/his work will look 'all wrong'. This is rarely the case. Ask questions to determine exactly what s/he is thinking. Then you can recognize the correct parts of her/his solution and pinpoint exactly what s/he misunderstands. To check each other's understanding you can create review questions to 'quiz' each other. In order to create a review quiz, you have to review all the material studied, decide which major concepts your group must understand before starting to learn new material, and create questions covering these topics. In order to make up a maths problem you have to know what kinds of questions make sense to ask and what information is needed to answer them. For example, it doesn't make sense to multiply the highest and lowest temperature in a given day; it does make sense to subtract to find the difference, or range, of those temperatures.

Two common misconceptions about group work are: 'I can't contribute anything to a group unless I first know how to solve the problem'; and, 'I am wasting my time in a group if I already understand how to solve the problem'. I learned about the first misconception when I asked students why they were each working separately even though they were sitting as a group as I had instructed them to solve the classwork problem together. Each person felt s/he couldn't contribute to the group effort until s/he could solve the problem alone. We then discussed out-of-school experiences we all had had where each individual in a group couldn't solve a particular problem, but by bouncing ideas off each other the group did solve the problem. We concluded this worked because different people understood different aspects of the problem, and because as each individual tried out a solution, others improved upon the ideas.[1] I answer the second misconception by arguing, as mathematics educator Julian Weissglass claims, that 'learning is facilitated by the learner communicating [his/her] knowledge. By explaining a concept to someone else, an individual increases [his/her] understanding. Many teachers report that they did not really understand a subject until they taught it' (1976, p. 17).

As Paulo Freire says, 'A project's method cannot be dichotomized from its content and objectives, as if methods were neutral and equally appropriate for liberation or domination' (1970b, p. 44); thus new methods, as well as content, are important in learning maths to understand and change the structure of society.

PROBLEMS

1 Read the following selection from Professor Julian Weissglass's maths-teacher preparation textbook (1979, pp. xiv-xv). Write about which of his points makes the most sense to you, and why. Also, write about one question you'd like to discuss that was suggested to you by his ideas. The question can involve such aspects as something he says that you think is wrong, something you feel needs further argument, or something you responded to so enthusiastically that it generated more ideas you'd like to explore.

'Learning in small groups from your peers is common, natural, and enjoyable. We learn games this way. To a great extent this is the way we learn language. If children began the study of language by first studying all the rules of grammar, language would be a very distasteful subject. Mathematics is simpler, more orderly, and easier to comprehend than a language such as English. Much of mathematics is governed by a small number of abstract rules. This tempts us to use the short cut of simply memorizing these rules rather than exploring and discovering mathematics in a concrete and human context.

'This short cut of memorization makes mathematics forbidding, deadens student interest, and can stand in the way of understanding. What do I mean by understanding here?

'*Understanding requires that new ideas and facts be related to those you already know and that all this information be remembered so that it can be used to solve old or new problems.*

'Notice that the understanding we will be seeking goes beyond mere memorization and includes the ability to generalize and to apply what has been learned in new contexts.

'Human beings are inherently curious and enjoy learning and understanding. Have you ever watched a child explore a puddle of water or investigate a leaf? Have you seen children's joy in building towers of blocks or their fascination with words or numbers? I have designed this book so that you will find the information and experiences in this course enjoyable, so that in turn you will be able to provide pleasant mathematical learning experiences for your students in the future. It is possible to provide classroom learning experiences that are both significant and enjoyable. There are two conditions that are essential to achieving these desirable results.

'First condition: *The attitude in the classroom must be positive.*

'If students are upset or disturbed or if they fear or dislike the subject, their full attention will not be available for learning. It is up to you, as a student and as a future teacher, to identify and deal with the fears, hesitations, and disturbances that get in the way of learning. While a class is not a therapy session, remember that from time to time we all experience anxiety about personal matters or classroom pressures. It is worth the effort to deal with these anxieties, both now with your classmates and in the future when you are teaching, because they destroy one of the preconditions for learning – a positive attitude toward the subject and the learning situation.

'Second condition: *Material must be presented in the proper context and at the proper rate for understanding.*

'If ideas are presented out of context or too rapidly, they are not understood. In addition, such an experience is distressing to the learner and thus violates our first condition. Furthermore, the effect does not stop there. We have all experienced situations in which we did not understand something, both in and out of school. Some of these early experiences were very hurtful. Although we may not identify them as such, they were *also* defeats in attempting to learn.

Many of the negative feelings and learning blocks that we have stem from these early experiences.

'Fortunately, it is not necessary to have completely recovered from the effects of previous distressing experiences in order to learn. We can create situations in which adequate attention is available for learning and proceed in ways that increase the likelihood that new information will be presented in context and at the proper rate. Although many lecturers are quite successful at this, I have found that the small-group laboratory approach has many advantages in seeking to satisfy the two conditions above.

'The small-group laboratory approach

'In the small-group approach, people learn from one another. This has benefits that are related to the conditions for learning. First, learning from peers eliminates the fear of authority that often interferes with learning. Children experience much criticism from parents and teachers. Because of these past experiences, even a teacher who does not intend to criticize students will still raise the fear of criticism in many students. For this reason it is easier to ask questions of or accept suggestions from a peer. Second, the communication of ideas that takes place in a small group is supportive of the learning process. The experience of helping someone learn will give confidence and overcome feelings of inadequacy that many people may have. Also, teaching others assists your own learning. One learns best what one is teaching. Finally, as a student, the small-group approach makes it easy for you to get to know your classmates. This will help you learn by reducing any feelings of isolation you may have. I encourage you to take advantage of this opportunity to get to know your classmates. One way of doing this is to take a few minutes to talk about something good that happened to you recently or about a good learning experience. Perhaps your instructor will suggest topics for you to talk about. Be sure that everyone gets a turn to speak and try to keep the discussion positive and cheerful. It will increase everyone's receptiveness to doing mathematics if attention is directed away from upsetting experiences and toward pleasant ones.

'The small-group laboratory approach also has definite advantages with regard to the second condition – that new information must be presented in context. An obvious difficulty with lectures is that of presenting information in the proper context and at the proper rate for a diverse audience. It is clear that more individualization than lecturing can provide may be needed to meet the second condition. I have found that groups of four work best, as they increase the resources available within each group yet still allow individual interaction.

'Asking questions

'Another advantage of a small group is that the student can assume a greater responsibility for the rate of presentation of new information. How fast and how well you learn depend largely on you. I think you will often be delighted and surprised and feel more confidence in yourself. However, sometimes you may feel frustrated or confused. This is also part of learning. The frustration and confusion can be reduced by the small groups themselves as they try to ensure that new ideas are presented in context and at the proper rate. One of the best ways to do this is by asking questions and encouraging others to do the same.

'Here are some suggestions for encouraging question-asking in your group:

'Do not criticize the questions of others nor hesitate to ask questions yourself for fear of criticism. Set an example for those in your group in this respect.

'Encourage questions by being friendly and cheerful and by expressing thanks and appreciation to those who ask questions.

'If someone is obviously confused, encourage that person to ask a question or ask a question yourself to try to clear things up.

'Use a series of simple questions to bring out an explanation rather than simply telling someone the answer.

'Remember that the *process* of learning is even more important than the *facts* learned.

'Show respect for your group and keep up morale by being on time and prepared. Being prepared means knowing what questions you want to ask; it does not mean knowing all the answers.

'Approach the group each day with aims of helping everyone else learn, asking interesting questions, and making the experience an enjoyable one.

'If your group encounters difficulties in freely asking questions, it is helpful to pair off and examine the causes of this problem and then report to the full group. You might each try to remember when you were able to ask questions freely and why it was that you stopped.

2 My students often feel that the positive aspects of group work are the opportunity to learn the different ways that others solve problems and the chance to gain improved self-confidence and understanding through helping others. On the other hand, they have often had negative experiences with people who dominated the group, or with feeling they had nothing to contribute to the group. I have found it helpful, then, to discuss the experience of educator David Reed, who worked with a group which 'challenged the notion of calling [some

people] the "silent ones" and affirmed that their form of participation . . . was a result of the dynamics of the entire group, not just individual problems . . . One conclusion of the discussion was that if the problems of some people not speaking up were to be resolved it would be through a collective effort including those who were more aggressive and vocal' (1981, p. 95).

List what you think are the advantages and disadvantages of learning maths with a group of people, and discuss why.

3(a) Formulate at least three questions about group work.

(b) Use these questions to interview three people about their experiences working in groups.

(c) Summarize their comments.

7 Multicultural considerations

Professor David Henderson, a white mathematician at Cornell University, argues that most mathematicians' views of mathematics are biased:

> Recently, I was thinking back over the times that my perception of mathematics had been changed by the insights or questioning of a person in my class. Suddenly, I realized that in almost all of those cases the other person was a woman or from a different culture than my own.
>
> Over the recent centuries the people in charge of mathematics, as we culturally define it, have been mostly Western (white), upper/middle-class males. So it should not be surprising if this has instilled a bias into our conception of mathematics . . .
>
> Evidence of this bias I see in the fact that most histories of mathematics downplay or ignore the role of non-Western cultures (for example, the Muslim culture during the West's 'Dark Ages'). There are mathematical ideas (such as 'Saccheri quadrilateral') which are named after the person (an Italian Christian) who first translated the idea into a Western language instead of being named after the person (Omar Khayyam, a Persian) who first introduced the ideas to the world (in Arabic). Though less often than in the past, women are still being told that they can't understand mathematics as well as men can. Working-class people are traditionally considered to not know mathematics; but I have seen from personal experience that a thinking carpenter or cabinet maker knows and uses a lot of geometry and understands it in a way that is different from, but just as correct as, what is normally taught in school . . .
>
> As I indicated above when I listen to how other people view mathematics my understanding of mathematics changes. I am certain that as women, and members of the working class and other cultures, participate more and more in the established mathematics, our societal conception of

> mathematics will change and our ways of perceiving our universe will expand. This will be liberating to us all. (1981, p. 13)

Not only are there methodological and conceptual differences in various cultures' mathematics, but there is also distorted and hidden history of these groups' contributions to mathematics. Based on your background, you may perform maths operations or perceive math concepts differently from the traditional school curriculum. By first understanding your mathematical point of view, you can understand other maths perceptions and then choose which (or create new) techniques and concepts that work best for you. Also, by realizing that people from all cultures have contributed to the development of mathematics, you can be inspired by various role-models in your learning of mathematics.

The methodological differences range from notation to ways of performing the operations. For example, in the United States and in Puerto Rico, twenty-three thousand is written with a comma – 23,000; in Haiti, in Vietnam, in Guatemala, and in Greece, twenty-three thousand is written with a period – 23.000. In the USA and in Puerto Rico, the period is used to separate whole numbers from decimal fractions, as in four dollars and twenty-nine cents – $4.29; in Haiti the comma is used to separate whole numbers from decimals as in their currency, 3 gourdes and twenty-five centimes – 3,25 G. As the maths topics are introduced in this book, many different methods will be presented. You can evaluate which make the most sense for you to use.

Some conceptual cultural differences stem from language. Linguists hypothesize that the structure of a person's first language has a significant influence on the cognitive processes involved in mathematics learning, such as classification and recognition of equivalences (Berry, 1985, p. 19). For example, in Setswana, the language of Botswana, a country east of the Kalahari desert in Southern Africa,

> Things are classified by what they *do*, rather than (as in our tradition which goes back to Plato), by what they *are* . . . If I point to an apple and ask 'Is this food?' (a more literal translation of the Setswana would be: 'Is this for eating?'), the question would be meaningless. To be more precise, the meaning would change from one context to another. If the apple were sitting on the table as part of a meal the answer would be affirmative; whereas if we were outdoors playing catch with the apple the answer would be 'no' . . . Contrast this with the usual classification, sorting and matching skills emphasized in the early stages of most 'new math' curricula. Children are required at the very beginning of their mathematics programme to master complex hierarchies of abstraction and classification whose schemata are derived from Indo-European models. We surmise that enormous tension between the child's natural classification modes and those required in the classroom is the inevitable consequence.

> One result is that mathematics . . . remains a mysterious subject . . . (Berry, 1985, p. 21)

Other conceptual learning differences may stem from Margaret Mead's observation that, 'Men and women – as they exist today, having been brought up in the present world – show marked differences in the way they tackle problems' (quoted in Rosamund, p. 4). Psychologists Carol Gilligan and Nona Lyons 'use the term "separate" to refer to traditional Western reasoning styles and the term "connected" to indicate the style they observed [in the women they studied]' (in Buerk, 1985, p. 63). Dorothy Buerk, a mathematics professor at Ithaca College, chose excerpts from Gilligan's book to characterize the two types of reasoning:

Separate Reasoning	*Connected Reasoning*
• gets right to solution in a structured, algorithmic way stripping away any context	• tries to experience the problem, relate it to personal world, clarify language, create context, remove ambiguity
• uses mode of thinking that is abstract and formal	• uses mode of thinking that is contextual and narrative
• geared to arriving at an objectively fair or just solution upon which all rational persons can agree	• geared to looking at limitations of any particular solution and describing the conflicts that remain
• legal elaboration of rules and fair procedures	• tolerant in attitude toward rules and more willing to make exceptions
• confident to judge	• reluctant to judge

> At a summer (1983) colloquium of the Mathematics Department of Ithaca College . . . I asked participants to tell me how each of these lists related to the teaching and learning of mathematics. Their basic consensus was that the 'connected' list represented the way that mathematicians do mathematics . . . And yet they agreed that the 'separate' list conveyed the way that mathematics is communicated in the classroom, in textbooks, and in their professional writing. For many, especially many women, this unfortunate disparity takes the life out of mathematics and mathematics out of their lives. (pp. 63–4)

In spite of these conceptual differences and the societal blocks of racism and sexism, people of colour and women have contributed to the development of mathematics. However, these contributions have been distorted and hidden. For example, in *The Number Concept*, L. L. Conant 'dismisses the amazingly complex numeration system of the Yoruba people of south-west Nigeria with these words:

> Nor on the other hand, is the development of a numeral system an infallible index of mental power, or of any real approach toward civilization. A continued use of the trading and bargaining faculties must and does result in a familiarity with numbers sufficient to enable savages to perform unexpected feats in reckoning. Among some of the West African tribes this has actually been found to be the case; and among the Yorubas of Abeokuta the extraordinary saying, 'You may seem very clever, but you can't tell nine times none', shows how surprisingly this faculty has been developed, considering the general level of savagery in which the tribe lived. (quoted in Zaslavsky, 1973, pp. 9–10)

Conant sees the occurrence of numbers up to a million among South African tribes as 'remarkable exceptions' to the 'law' that 'the growth of the number sense keeps pace with the growth of the intelligence in other respects'.

Regarding the non-Western origins of mathematics, Beatrice Lumpkin writes (1983) about how, despite the fact that

> no pictures have come down to us of any of the great men and women of Alexandria, false portraits have been published which portray them as fair Greeks, not even sunburned by the Egyptian sun . . . In the case of Euclid, best known of the Alexandrian mathematicians, there is not a shred of evidence to suggest that he was anything other than Egyptian . . . [and] it is highly probable that [he was] African. (pp. 104–5)

For another example of hiding contributions,

> when the Eiffel Tower was erected, in which the engineers were obliged to give special attention to the elasticity of the materials used [an 11.8-inch scale model of it weighs only 7 g or $\frac{1}{4}$ oz!], there were inscribed on this lofty structure the names of seventy-two savants. But one will not find in this list the daughter of genius, whose researches contributed so much toward establishing the theory of the elasticity of metals – Sophie Germain. (Osen, 1974, p. 92)

Unfortunately, race and sex discrimination continue to put hurdles in the way of full participation for people of colour and women. For example, in 1950, Evelyn Boyd Granville, a black woman who earned a PhD in mathematics at Yale, applied for an academic position: 'When the hiring committee at one institution discovered she was Black, they merely laughed at her application and never considered her for the job.' For another example, although black scholar Elayne Arrington-Idowu finished first in her class in an integrated high school in Pittsburgh, she was not allowed to give the valedictory address. 'A scholarship she had won to attend the University of Pittsburgh was withdrawn, in this case because she was female.' But the struggle is not over. She is now a mathematics professor at that university, and increasing numbers of people of colour and women are becoming mathematicians and mathematics educators (Zaslavsky, 1983, p. 111).

For more on multicultural considerations, see Hemmings (1980), Joseph (1987) and Zaslavsky (1973).

PROBLEMS

1 At one time Western anthropologists claimed that traditional non-literate cultures were 'childlike' and 'primitive'. Marcia Ascher, a mathematician, and Robert Ascher, an anthropologist, argue that 'there is not one instance of a study or a restudy that upon close examination supports the myth of the childlike primitive' (1986, p. 131). They quote Cole, Gay, Glick and Sharp (p. 233) to conclude that 'cultural differences in cognition reside more in the situations to which particular cognitive processes are applied than in the existence of a process in one cultural group and its absence in another'. What follows are three examples the Aschers give to illustrate Westerners' misunderstandings and misrepresentation of mathematics in other cultures. In your own words, discuss why these examples support the Aschers' argument, and speculate about why you think the early anthropologists concluded that traditional non-literate peoples were 'primitive'.

– '[A frequently repeated] anecdote tells of an exchange between a native African Demara sheep herder and someone else variously described as an explorer, trader, scientist . . . It is intended to show that the herder cannot comprehend the simple arithmetic fact that 2 + 2 (or 2 × 2) = 4. It describes how the herder agrees to accept two sticks of tobacco for one sheep but becomes confused and upset when given four sticks of tobacco after a second sheep is selected. Of course, the problem is not that the shepherd doesn't understand arithmetic; it is rather that the scientist/trader doesn't understand sheep. Sheep are not standardized units. [The Demara herder's] confusion could be attributed to the trader's willingness to pay an equal amount for the second, different animal.' (p. 128)

– [Researcher Lévy-Brühl noted] that certain numbers and numbers that are multiples or divisors of them play an important role in Vedic religion, rituals, and legends. In this context, on occasion 3, 7 or 9 are substituted for each other. From this Lévy-Brühl concludes 'this equivalence, an absurdity to logical thought, seems quite natural to pre-logical mentality, for the latter, preoccupied with the mystic participation, does not regard these numbers in abstract relation to other numbers, or with respect to the arithmetical laws in which they originate'. But a recent field study of the Kédang, who also have this type of substitution, enables a different conclusion . . . when used in symbolic contexts, odd numbers are associated with life and even numbers with death. Substitutions within these classes are possible if circumstances require it. If, for example, a ceremonial period of 4 days is stipulated but cannot be met, 2 days will do but 3 would be a serious infringement . . . The formation of these equivalence classes is, we think, instead an example of an abstract idea about numbers.' (pp. 129–30)

– 'The use of syllogisms for investigating the reasoning ability of non-literate people was recently re-examined by cognitive psychologists. The study probed more deeply, reasoning that respondents used to arrive at what were formally viewed as unsatisfactory responses. Here is one example in which a Kpelle respondent would not reply to the question.

Question: All Kpelle men are rice farmers. Mr Smith is not a rice farmer.
Is he a Kpelle man?

One part of the response: If you know a person, if a question comes up about him you are able to answer. But if you do not know the person, if a question comes up about him, it's hard for you to answer.

Although the syllogism posed by the questioner has gone unanswered, the ability to reason and to think hypothetically has been demonstrated. What has also been demonstrated is that the Kpelle respondent and his western questioner have different views on talking about people whom you do not know. (p. 130)

2 Read the following excerpt from an article (in *International Education*) written by Munir Fasheh, Dean of Students at Palestine's Birzeit University, while he was studying for his doctorate at the Harvard Graduate School of Education.

(a) Why was his mother's knowledge of mathematics hidden?

(b) Write about what kinds of mathematical knowledge are similarly 'hidden' in your family or community.

When structures fall people rise

'When I returned to Birzeit in 1971, I was filled with energy in two different directions: the one, to expand the use of logic and science in the world through teaching, and the other, to deal with what we experienced as an attempt to dismantle the Palestinian community as a viable entity. Opportunities in mathematics presented themselves almost immediately. While the Arab countries had already introduced the 'New Math', the West Bank and the Gaza Strip, being under military occupation, had been left out. Birzeit organized a course for all high school teachers in the West Bank in the summer of 1972. I ran that program and helped to incorporate cultural concepts, independent exploration, and affective engagement into the syllabus, to overcome the fundamentally dry and alien abstraction of the math. Both teachers and students were enthusiastic about this revitalization of the teaching but it did not yet lead me to question hegemonic assumptions behind the math itself.

'The Palestinian community I went back to was self-confident, energized, idealistic, and already involved in its own renewal, largely as the result of the development of the Palestinian movement. A group of us began children's programs in drama, art, crafts, mathematical games, simple science experiments, poetry, music and literature, which developed and expanded quickly. We also began working voluntarily in other community projects. While these activities in the community involved joyousness, spontaneity, cooperation and freedom, they were not yet fully articulated for me as education and were not yet fully praxis in Freire's sense.

'While I was using mathematics to help empower other people and while I was being empowered by the voluntary work, mathematics itself was not empowering me. It was, however, for my mother, whose theoretical awareness of mathematics was completely undeveloped. Math was necessary for her in a much more profound and real sense than it was for me. My illiterate mother routinely took rectangles of fabric and, with few measurements and no patterns, cut them and turned them into beautiful, perfectly fitted clothing for people. In 1976 it struck me that the mathematics she was using was beyond my comprehension; moreover, while mathematics for me was a subject matter I studied and taught, for her it was basic to the operation of her understanding. What kept her craft from being fully a praxis (in Freire's term), and what limited her empowerment was a social context which discredited her as a woman and uneducated and paid her extremely poorly for her work. Like most of us, she never understood that social context and was vulnerable to its hegemonic assertions. She never wanted any of her children to learn her profession; instead, she and my father worked very hard to see that we were educated and did not work with our hands. It was a shock to me to realize, in the face of this, the complexity and richness of her relationship to mathematics. Mathematics was integrated into her world as it never was into mine.

'My mother's sewing demonstrated another way of conceptualizing and doing mathematics, another kind of knowledge, and its place in the world. The value of my mother's tradition, of her kind of mathematics and knowledge, while not intrinsically disempowering, however, was continuously discredited by the world around her, by the culture of silence and cultural hegemony.

'The discovery of my mother's math was a discovery about the world and the relations of hegemony and knowledge. Hegemony does not simply provide knowledge; it substitutes one kind of knowledge for another in the context of a power relationship. While I had been struggling to make the mathematics I had learned meaningful, the embodiment of what I was seeking was in front of me, made invisible to both my mother and me by the education I had been given, which she desired for me. It had been, in Freire's term, an education for oppression, domestication, and dehumanization. While I was not yet ready to question the theoretical bases of Western science and math themselves, the discovery allowed me to recognize the need for liberated education, to respect all forms of knowledge and their relation to action.'

8 Self-evaluation

Most basic maths programs involve standardized testing. The history of these kinds of tests indicates that they are filled with difficulties, from bias to confusion about the meaning of the results. For example, when the Stanford-Binet IQ test was first prepared in 1916, women had lower scores than men; only in 1937 was the test redesigned so that men and women had the same mean score. But the cultural bias in favour of whites has still not been corrected (Beckwith and Durkin, 1981, p. 8). For another example, the US Army Mental Tests, which were supposed to measure 'native intellectual ability', included such items as

- Crisco is a: patent medicine, disinfectant, toothpaste, food product.
- The number of a kaffir's legs is: 2, 4, 6, 8 (Gould, 1981, pp. 199–200).

The second question is completely ridiculous, because 'Kaffir with a capital K is a Bantu-speaking South African, while kaffir with a lower case k is a form of sorghum' (Seldon, 1983, p. 179).

However, there are constructive ways of evaluating your own work that can be very useful learning experiences. A colleague from Rutgers University in New Jersey, Arthur Powell, gets students to complete the following self-evaluation, and then meets with them to discuss their comments along with his evaluation of their work, before giving out grades.

Self-evaluation

Criteria

Daily assignments

Keep up to date

Get back assignments when absent

Seek assistance when needed

Long-term assignments

Work at a reasonably even pace
Look for patterns
Work independently
Class participation
Listen to other students
Be prepared with daily assignments
Ask questions
Perseverance
Use all available out-of-class resources
(tutors, friends, instructors)
Take full responsibility for absences
(obtain notes and assignments from class-mates)
Work at highest level of capability
Progress as a mathematician
Understand structures of mathematics
Use patterns to increase understanding
Utilize new or improved techniques of calculation

Evaluation (attach a separate sheet)
Areas of strength:
Areas needing improvement:
Possible grades: A, B+, B, C+, C, D, F ‹Average grade C+›
Grade:
Rationale for grade: (Use other side.)

By keeping track of your work this way, you can figure out the kinds of things you need to do to learn more effectively.

Another kind of self-evaluation involves working on actual maths problems. As we discussed before in Chapter 1, it is a misconception that self-evaluation is cheating and that 'expert' teachers are the only ones who can evaluate your work. In the selfevaluation at the end of this chapter, I recommend that you check the answers after you finish each exercise. By trying to analyse what you did correctly and *exactly* what you did incorrectly, you can learn from your mistakes and apply this learning to solving the next problems. The problems are ones you'll be able to solve when you *finish* both volumes of this text. So, you may get every answer wrong. This does not mean you do not know any maths. There is always some correct reasoning involved in any attempt to solve a problem. Self-evaluation is an important process in your taking control of your learning by focusing on what you currently understand and what questions you want answered to increase your understanding.

The following are two examples, one very basic and the other quite complicated, that illustrate how wrong answers can involve some correct reasoning, and that discuss ideas about how to start solving problems and assess where you get stuck.

Example 1: Someone who does the following series of additions

$$\begin{array}{r} 46 \\ +29 \\ \hline 65 \end{array} \qquad \begin{array}{r} 38 \\ +54 \\ \hline 82 \end{array} \qquad \begin{array}{r} 128 \\ +\ 75 \\ \hline 193 \end{array}$$

knows the basic addition facts (for example, 6 + 9 = 15) and understands that each place-value column is added separately, but needs to review 'carrying'.

Example 2: Someone who doesn't see the entire solution to the following application problem can still understand some aspects of it and begin to work on the solution.

> The United States has about 9,000 strategic nuclear warheads pointed at the Soviet Union. The Soviet Union has about 6,000 strategic nuclear warheads pointed at the United States. The combined population of the United States and the Soviet Union is about 500 million. Each warhead could kill about 2 million people. If all these warheads were released, about how many people could be killed?
>
> (These statistics are from an article in the *Nation*, 25 August 1981, reviewing a CBS Reports series on 'The Defense of the United States'.)

To begin, slowly read the problem two or three times. Try to rewrite the problem in your own words, temporarily ignoring the numbers, stating what the situation is and what you are asked to find. One way is: The USA and the USSR have loads of nuclear warheads. Each one can kill a tremendous number of people. What is the total number of people that all these warheads could kill? Once you understand the situation, you can go back and pick out which numbers you need. If you get stuck here, it does not mean that you are 'stupid' or that you should give up! There are many numbers given, some of which have to be combined (sooner or later), one of which is not needed in the solution. When you get stuck, don't spend too long at one sitting grappling with the problem. It's a misconception that maths is done by working intensely until the problem is solved; going away from a problem and later returning to it allows your mind to assimilate ideas and develop new ones. 'Resting' is not giving up. We all have different frustration levels; with experience you'll learn the best time for you to rest from work on a particular problem.

Coming back to the problem you may see that you can add to find the total number of warheads in both countries (9,000 + 6,000 = 15,000 warheads), and since each warhead kills 2 million people, you can multiply 15,000 × 2 million to find the total number of people who could be killed (15,000 × 2,000,000 = 30,000,000,000 = 30 billion people). Or you may see the equally correct method of finding out how many people each country could kill (US: 9,000 × 2 million; USSR: 6,000 × 2 million) and then adding to find the total. (This is also an illustration of how there are often many correct ways to solve each maths problem.)

If your second try at the problem is unsuccessful, find someone with whom you can work. This is not cheating! We learn best by asking questions, listening to others' ideas, evaluating which ideas make sense to us, analysing why we got stuck, and learning from our mistakes. In this problem you may have been confused by the 500 million total population figure, which is not needed to answer the question (although it is needed to understand the concept of overkill: since the weapons can kill 30 billion people, the USA and the USSR could kill each other 30,000,000,000 ÷ 500,000,000 = 60 times over. (The concept of overkill, therefore, supports arguments that we should reallocate tax money from the military to public and community service.) Or, you may have got stuck because you were unclear about the meaning of multiplication which enables us to find a total when each group we are totalling has the same number (each nuclear warhead kills 2 million people). Or, you might have understood that you needed to multiply 15,000 × 2 million, but you didn't know how to multiply such giant numbers. Or, you might have become so upset and angry at the information given in the problem, that you didn't want to proceed further . . . and needed to talk with the person who created the evaluation about why s/he feels that critically analysing the distressing facts about our world is an important step in changing those facts. Or, you may have got stuck for any of many other individual reasons.

The self-evaluation that follows gives you an overview of the range of problems you will be able to solve *after* you complete both volumes of this text. The answer section includes questions which suggest some of the key aspects involved in the solution of each problem, to help you evaluate where you might have made an error. You are *not* expected to be able to solve any of these problems *now*. The important skills to work on throughout your maths learning are: to know exactly what you do understand about a problem; to learn a language to express precisely the questions you need answered in order to solve the problem; and, to develop resources (the teacher, other students, books, and so on) to help you with your studies.

Further, to help you evaluate your work, this text contains answers to all problems and, throughout, you are asked to create and solve your own maths problems and review quizzes. In this way, you learn more completely, because in order to create your own evaluations you need to understand all the individual maths concepts and their interrelationships and which concepts are the major ones that must be understood before learning new material. You also need to know what kinds of questions make sense to ask and what information you must include so that the problem can be solved. In this text you will learn techniques for creating good questions, and you will be asked to evaluate my questions, discussing whether and why they are hard or easy; interesting or boring; relevant or irrelevant; and so on. Whenever you are learning anything, I suggest not only evaluating what you do and don't understand, but why, whether you are bored, challenged, angry, etc.

PROBLEMS

Self-evaluation

Remember These problems represent an overview of the topics we will cover in both volumes of this text. You are not expected to be able to solve these problems now. But you are expected to try each problem and identify, as precisely as you can, everything about it you understand and where you get stuck.

Drill

(Answers at end of chapter)

1 $35.04 - 3.428 =$

2 $12\frac{1}{4} - 9\frac{5}{6} =$

3 $3.72 \times 0.008 =$

4 $2\frac{4}{5} \div 1\frac{1}{3} =$

5 Rewrite in decimal form: (a) $\frac{3}{7} =$
(b) $3.6\% =$
(c) $12\frac{3}{4}\% =$

6 Rewrite in fraction form: (a) $0.063 =$
(b) $250\% =$
(c) $4\frac{1}{2}\% =$

7 Rewrite in percentage form: (a) $\frac{5}{8} =$
(b) $0.005 =$
(c) $5\frac{3}{5} =$

8 Read the following article (*In These Times*, 1 February 1984).

Reading the colour chart

'While the Commission on Civil Rights was busily devising ways to protect whites from discrimination, the Chicago Urban League published a report last November that reinforced the "unenlightened" notion that to be black or Hispanic in a major US city still means less income, less chances to finish high school and a greater probability of unemployment. While no surprise to anyone (perhaps even Linda Chavez *et al.*), the statistics painted the income disparity between these minority groups and whites in sharp relief. At $12, 716, the average black income in Chicago is half that of the average white. San Francisco-Oakland and Washington, DC, follow closely on

Chicago's heels with income differences in the $11,000 range. Although in general Hispanics fare slightly better than blacks in the income category, in Boston they have the lowest average income ($9,586) and the highest differential ($13,661). Apparently life in the Northeast is especially cruel for Hispanics, with New York and Philadelphia capturing second and third place in the income gap.'

(a) Find the average white income in Chicago.
(b) Find the average white income in Boston.
(c) Construct a table presenting the data in the article.

9 Use the following chart from *Laurel's Kitchen* cookbook to find two different ways for an adult woman to meet the suggested daily requirement of iron.

Sources of iron (partial listing)

Recommended daily allowance for adults is 10–18 mg
Recommended daily allowance for adult women is 18 mg

Prune juice, 1 cup	10.5 mg
Black beans, 1 cup, cooked	7.9 mg
Garbanzo beans, 1 cup, cooked	6.9 mg
Lentils, 1 cup, cooked	4.2 mg
Peach halves, dried, 5	3.9 mg
Split peas, green, 1 cup, cooked	3.4 mg
Raisins, $\frac{1}{2}$ cup	2.6 mg
Tofu, 4-oz piece	2.3 mg
Tomato juice, 1 cup	2.2 mg
Wheat germ, $\frac{1}{4}$ cup	1.9 mg
Acorn squash, $\frac{1}{2}$ baked	1.7 mg
Strawberries, 1 cup	1.5 mg
Potato, cooked, large	1.4 mg

10 In 'Another day, another $3000: executive salaries in America', Paul Blumberg examines the range of income inequality in our society which he says 'extols its egalitarian tradition and during periodic bouts of national self-delusion even proclaims its essential classlessness' (p. 316). In 1976, the median income of all full-time blue-collar working males was $12,500 before taxes. In 1975, the chairman of Rapid-American Corporation paid himself (he, in essence, determines his own salary) a salary of $366,000 and a bonus of $550,000. What was his weekly gross pay?

11 Ocean sun-fish lay eggs that hatch into fish only $\frac{1}{10}$ in. long. These baby fish grow to about 960 times their birth length (Bell, p. 34). How long is an adult ocean sun-fish?

12 The birth of young bear cubs is during the winter sleep of bears; they

sometimes weigh no more than $\frac{1}{2}$ lb. An adult bear may weigh about 200 lb (Bell, p. 35). How many times greater is the adult bear's weight than the cub's weight?

13 According to an article in the *New York Times* (8 April 1979), 'despite the high cost of white-collar crime, the Justice Department spent only 5.1 percent of its total resources to combat such crime over the past two years . . . ' The General Accounting Office estimated that white-collar crime cost the American public $200 billion annually. Street crimes involving property cost the public 50 times less. How much money do Americans lose annually from street crimes involving property?

14 The unpaid labour done largely by women at home is an important factor in increasing business profits – wages would have to be much greater if men had to pay people to cook, clean, shop, and so on for them. Of course, there has been a tremendous increase in the participation of women in wage labour outside the home: over half of all married women with school-age children are wage-workers, as are over a third of married women with pre-school-age children. Yet the amount of time women spend working in the home is still substantial. Use the following chart to create and answer a maths problem whose solution involves subtraction and multiplication.

Average daily time spent on household work

Number of children	Age of wife or youngest child	Non-employed-wife families		Employed-wife families	
(1)	(2)	(3a)	(3b)	(4a)	(4b)
	Wife	*Wife*	*Husband*	*Wife*	*Husband*
None	Under 25	5.1	0.9	3.5	1.4
	25–39	5.9	1.2	3.6	1.4
	40–54	6.2	1.5	4.3	1.8
	55 and over	5.4	2.0	4.3	1.1
	Youngest child				
One or more	12–17	7.1	1.7	4.8	1.7
	6–11	7.4	1.6	5.4	1.5
	2–5	8.2	1.6	6.2	1.7
	1	8.8	1.7	6.2	3.5
	Under 1	9.5	1.5	7.7	1.6

How to read this table:

Take, for example, a family with no children (col. 1) where the wife is between 25 and 39 years of age (col. 2 and row 2). If the wife is not employed (col. 3) she does an average of 5.9 hours per day (col. 3a) while her husband does an average of 1.2 hours per day (col. 3b). Even if the wife in this family works (col. 4) she still does 3.6 hours of housework (col. 4a) per day while her husband does an average of 1.4 hours per day (col. 4b).

Source: *Handbook on Women Workers 1975*, based on a sample of 1,378 families in upstate New York in 1967–8 and 1971.

To pursue this topic, see Edwards, Reich and Weisskopf (1978).

15 Use the line graph that follows to estimate the difference between white and black and other families' median income in 1973.

Median annual money income of families, by race, in constant 1976 dollars: 1960 to 1976

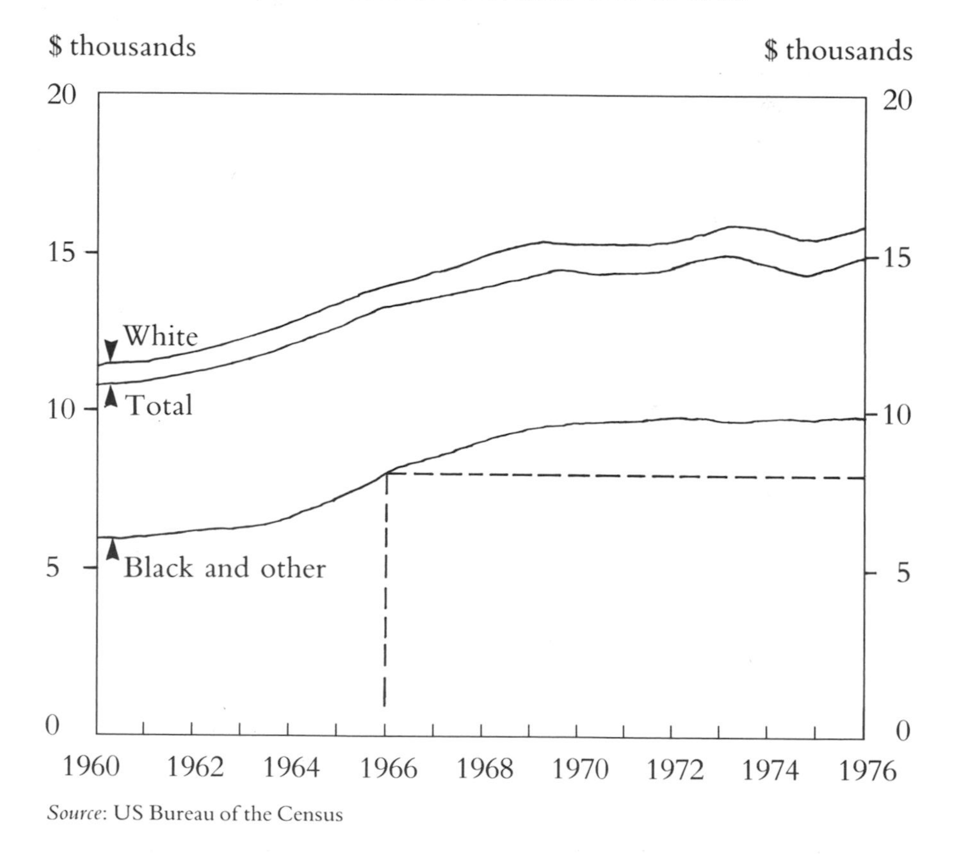

Source: US Bureau of the Census

16 If you keep the number of calories you eat constant, exercise can contribute to weight loss. For example, fast bicycling (at a speed of 13 miles per hour) burns up 0.072 calories per minute per pound of your weight. Round off this measure to the nearest hundredth of a

calorie and estimate how many calories a 200-lb person would burn up by bicycling fast for half an hour.

17 A June 1981 article in *The Progressive* reports that in 1981 MIT received a university-military research contract for $154,564,000. (This was the second largest out of hundreds of such contracts.) Round this number to the nearest million and describe how to estimate the number of students who could have attended MIT for free if our tax money had gone to awarding scholarships instead of for military research.

18 In 1983 the richest $\frac{1}{200}$th of the US population owned 35.1 percent of all the wealth (*Boston Globe*, 26 July 1986). What percentage of the population owns 35.1 percent of all the wealth in the USA?

19 The A. C. Nielsen Company estimated that 12.9 percent of the 28,840,000 people watching prime-time situation comedy TV from October to December 1977 were in their teens (aged 12–17). How many teenagers did Nielsen report watched situation comedies during that time period?

20(a) In 1984, what percentage of dentists were men?
(b) In 1984, what percentage of physicians' assistants were black?
(c) In 1984, how many dentists were non-black?

Employed persons* in selected health occupations: 1984

Occupation	Number (1,000)	Percentage Female	Percentage Black
Physicians	520	16.0	5.0
Dentists	138	6.2	0.9
Registered nurses	1,402	96.0	7.6
Pharmacists	162	28.5	2.9
Dieticians	70	95.1	16.1
Physicians' assistants	43	37.2	3.3
Managers, medicine and health	96	61.2	6.4

* Covers civilians sixteen years old and over.

Source: Statistical Abstracts of the United States 1986, No. 161, p. 103.

ANSWERS: CHAPTER 8

First, remember that you are not expected to be able to solve any of these problems until you complete both volumes of this text. Of course, some of you will remember some mathematics and may solve some of these problems now. For each of your incorrect answers, try to identify your specific mistake *and* identify at least one thing that you did correctly. Following each answer is a series of questions suggesting some of the key aspects of a solution to each problem. However, the questions do not cover all aspects of every method of solution; they are just intended to help you start a specific evaluation of your maths knowledge. Also, if your answer is different from the answer given here, it does not necessarily mean that you are wrong. As we discussed in Chapter 1, a misconception about mathematics is that there is only one answer to every problem. On the contrary, your answer to an application problem can depend on what assumptions you make related to the given information – some problems are open to a number of different interpretations. So if you have a different answer from the one on this key, and after considering the suggestions listed and reviewing your own solution, you still feel your answer is correct, discuss it with someone until you determine whether or not both answers are correct.

1 31.612. If you used paper and pencil, did you write 35.04 as 35.040, so you could subtract the eight thousandths in 3.428 *from* zero thousandths in 35.040?

2 *Paper and pencil*: $2\frac{10}{24} = 2\frac{5}{12}$
Did you find a common denominator?
Did you rewrite the 1 you borrowed from the 12 as $\frac{24}{24}$, a fraction with the common denominator?
Calculator: 2.42 (rounded to the nearest hundredth)
Did you divide the denominator *into* the numerator (for example, 5 ÷ 6) to change the fractions to decimals?
If you used the memory, did you put the $9\frac{5}{6} = 5 \div 6 + 9$ into [M−] ?

3 0.0296. If you used paper and pencil, did you count the number of decimal-fraction places in *both* factors to determine the number of decimal-fraction places in the answer?
Did you make the fifth place by adding a zero to the left of the digits obtained from multiplying 372 × 8?

4 *Paper and pencil*: $2\frac{1}{10}$
Did you invert the second fraction and then multiply?
Calculator: 2.1 (rounded to the nearest tenth)
If you used the memory, did you put the second fraction ($1\frac{1}{3} = 1 \div 3 + 1$) into the memory *first* so that you could divide in the correct order?

5(a) 0.43 (rounded to the nearest hundredth)
(b) 0.036
(c) 0.1275. After rewriting $\frac{3}{4}$ as 0.75, did you then divide the 12.75 by 100 in order to rewrite the % as a decimal?

6(a) $\frac{63}{1,000}$
(b) $2\frac{1}{2} = 2.50 = 2.5$. If you used the short cut for dividing by 100 (moving the decimal point two places to the left), did you start the decimal point of the whole number 250 after the ones' place (that is, 250.)?
(c) $\frac{45}{1,000}$. After rewriting $4\frac{1}{2}$% as 0.045, did you then rewrite the decimal as a fraction?

7(a) 62.5%
(b) 0.5%

(c) 560% After rewriting $5\frac{3}{5}$ as 5.6, did you then multiply by 100 to rewrite the decimal as a percentage?

8(a) $12,716 × 2 = $25,432. Did you divide by 2 (because of the word 'half') rather than multiply? ('Black income = half white income' means the same as 'white income = twice black income'.)

(b) $9,586 + $13,661 = $23,247. Did you subtract instead of add because of the word 'differential'?

(c)

Group	Chicago	Boston
Black	12,716	
Hispanic		9,586
White	25,432	23,247

Only a partial table can be created. Did you get confused because you cannot fill in the entire table from the information in the article?

9(a) 1 cup of prune juice + 1 cup black beans = 10.5 + 7.9 mg = 18.4 mg

(b) 3 large potatoes + 1 acorn squash + 8 oz tofu + 1 cup garbanzos = 3(1.4) + 2(1.7) + 2(2.3) + 6.9 = 19.1 mg.

Did you interpret the question to mean the sum had to equal *exactly* 18 mg?

10

$$\frac{366,000}{52} = \$7,038.46$$

Did you divide in order to find out the size of each weekly piece, given the total salary and number of pieces?

Did you add the bonus and the salary first before dividing by 52? This is certainly a reasonable assumption, although the weekly pay cheque is probably figured from the salary, and the bonus given in one separate cheque. So, we would also consider as a correct answer:

$$\frac{366,000 + 550,000}{52} = \$17,615.38.$$

Did you ignore the $12,500 median income of full-time blue-collar working males, which is not needed to solve this problem? (Although it is certainly relevant in a comparison of the economic inequality in the United States.)

11 $960 \times \frac{1}{10}$ = 96 inches ($\frac{96}{12}$ = 8 feet long!)

12 $200 \div \frac{1}{2}$ = 400 times greater. Even though the word 'times' is used in this problem, did you divide to find out *how many times greater* (that is, the baby starts at $\frac{1}{2}$ lb and increases in weight 400 times, $\frac{1}{2} \times 400$, to 200 pounds as an adult)?

13

$$\frac{200 \text{ billion}}{50} = \$4 \text{ billion}.$$

Did you ignore the 5.1 percent, which is not needed to solve this problem?

14 One of the many possible problems: how much more work per week does an employed wife do than her husband (when their youngest child is two to five years old)? Answer: 7(6.2) − 7(1.7) = 31.5 hours.

15 $16,000 − $10,000 = $6,000. Did you round off to the nearest thousand? (If you rounded to the nearest five thousand, your answer of $15,000 − $10,000 = $5,000 would also be correct because the problem does not indicate to which number you should round.)

16 0.07 × 30 × 200 = 420 calories. Did you locate the 7 as the hundredths' place and then round 0.072 to 0.07 because the 2 in the thousandths' place is less than 5?

Did you translate 'per minute' to 'for each minute' and therefore multiply 0.07 by the 30 minutes?

(By the way, 1 lb = 3,500 calories, so if a 200-lb person rode a bike fast for half an hour a day for two weeks, s/he would lose 420 × 14 = 5,880 calories, or almost $1\frac{3}{4}$ pounds ($\frac{3,500}{5,880} = 1.68$).

17(a) $155,000,000. Did you locate the millions' place and round up one million since the number in the hundred thousands' place is 5 or greater?

(b) First find out the tuition fees at MIT. Then divide the $155,000,000 by the tuition fees to find out how many tuitions could have been paid with the $155 million military-research contract.

18 0.5% (or one half of 1%). Note that the comparison is more meaningful when both figures are in the form of percentages.

19 28,840,000 × 0.129 = 3,720,360. Did you translate 'of' to the operation of multiplication?

Did you rewrite the percentage as a decimal before multiplying?

20(a) 100% − 6.2% = 93.8%. Did you find the data from the percentage column, not from the numbers column?

Did you subtract the given percentage from 100% since the categories on the chart are the complements of the percentage asked for (that is, % women + % men = 100%; % black and other + % white = 100%).

(b) 3.3%. Did you subtract here, when you can just read the number from the chart?

(c) 0.9% of dentists were black, so 100% − 0.9% = 99.1% were white or Asian, etc. 0.991 × 138,000 = 136,758 non-black dentists.
Did you find the percentage of white or Asian, etc. dentists by subtracting the percentage of black dentists from 100%?
Did you know that to find a percentage of a number you multiply?
Did you realize the number in the number column must be multiplied by 1,000?

Part Two

The meaning of numbers and variables

a	b	c	d	e	f	g	h	i	j	k	l	m	n	o	p	q	r	s	t	u	v	w	x	y	z
1	2	3	4	5	6	7	8	9	10	11	12	13	14	15	16	17	18	19	20	21	22	23	24	25	26

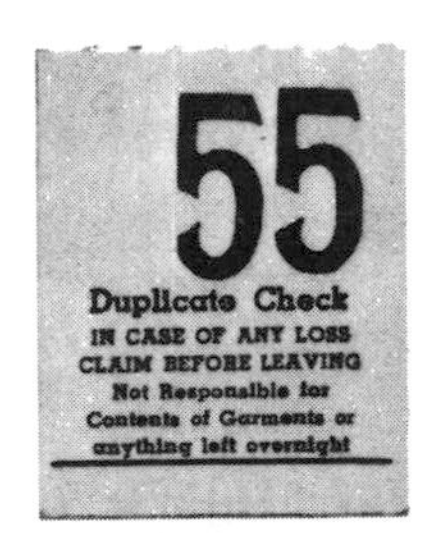

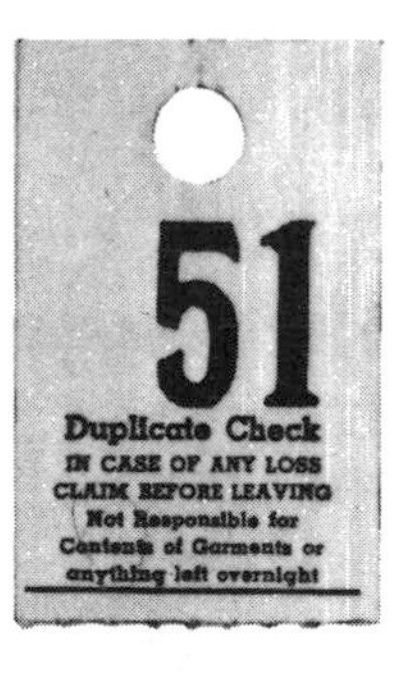

9 The meaning of whole numbers

I vividly remember a meeting I had eighteen years ago with the assistant principal of a school at which I was teaching. It was my first work experience, and I was questioning my boss. Why did we have to teach the students about roman numerals? He looked at me condescendingly, and replied that learning the roman system helps students to understand the usefulness of the decimal, place-value system for writing and computing with numbers. I could not accept this answer, which had no connection with the reality of our lives in the school. The students thought the entire maths curriculum was meaningless, useless, and boring. They certainly had no interest in using, let alone understanding, the place-value system! The assistant principal would not listen to my analysis of the situation, and at that time I had no better suggestions of what maths to teach, so the discussion ended. I initiated many similar discussions that year, not only about maths, but also about the many, many rules teachers and students had to follow just because 'that's the way we do things here'. The response was always the same. Finally, I could no longer continue teaching useless facts and obeying meaningless rules, so I quit. Amazingly enough, no one would believe why I had quit; they were all sure I must be pregnant.

Since then I have worked in a variety of alternative, public high schools and colleges. I have learned a great deal about teaching through my experiences with students and dedicated teachers. Over the last few years, I have learned a great deal about the relationship between education and social change through reading, and through reflecting about my experiences. Now I can think of an important reason to know that there are different ways of writing numbers. A cultural history of numbers makes clear that they are not mystical symbols handed down to us by mythical beings. Numbers are created by human beings and different number systems have developed over time in response to the circumstances of people's lives. Because number systems have been created by human beings, they can be re-created or changed by human beings to meet the differing needs of more and more complex societies. Statisticians Shaw and Miles hypothesize that in a society dedicated to justice and human liberation

> We would replace accountancy in terms of money and profit by accountancy in terms of social needs. We would replace the definition of social goals by those at the tops of the bureaucratic pyramids, by democratic self-control over all collective activities. We would then require new ways of measuring our needs and goals which expressed their great variety rather than reduced them to money values or standards imposed from above. (1979, p. 36)

Perhaps, if the cultural history of political, economic and social systems was presented as the history of continually changing institutions created by human beings to mediate their conflicting needs, we would not feel so powerless to change the systems under which we are currently living.

When I first started writing this section, I knew very little about the history of number systems. About all I knew was that the early systems of writing numbers, such as the roman system, were clumsier to write and much harder to compute with. Roman numerals use a different symbol for each 'grouping' (for example, C stands for 100, X stands for 10, and I stands for 1) and represent numbers by lining up, in order, as many of each grouping as needed. For example, 234 may be written in the roman system as CCXXXIIII. One of the first place-value systems, developed by the Chinese, created separate symbols to represent how many of each grouping were in the number. In a system like the Chinese, 234 would be written as 2C, 3X, 4I (Menninger, pp. 53–9). The decimal place-value system we use, called Hindu-Arabic, originated in India about a thousand years ago and was spread to Europe by North Africans of Muslim culture. It only uses the digits and represents the groupings abstractly, by their positions. So the '2' in 234 means 'two hundreds' (or 2C, or CC), because the third place, counting from right to left in whole numbers, represents hundreds. Similarly, the '3' in 234 means 'three tens' because the second place means tens.

There is something fascinating that I didn't know when I started writing this section: despite the fact that the Mayans of Central America used the zero hundreds of years before its earliest known use in India in 876 AD (Lumpkin, 1983), the concept of the digit zero was one of the key intellectual obstacles to adopting Hindu-Arabic numerals in Europe. 'The zero is something that must be there in order to say that no thing is there . . . For the primitive reckoner number is always a *number*, a quantity, and only a number can have a symbol' (Menninger, 1969, p. 400). The only European group which quickly and gladly adopted the new numerals was the astrologers whose status was increased by knowledge of this secret form of writing.

> . . . the new numerals were adopted in the early Middle Ages not because of any conception of the advantages of place-value notation but merely as a new and exotic means of writing numbers . . . the place-value notation gradually came to be understood . . . but this would have happened much more slowly than it did were it not for a strong external stimulus. This stimulus was provided not by the scholars and scientists, but by the merchants. (Menninger, 1969, pp. 423–4)

An interesting example of how resistant societies become to change is that, in spite of the fact that Hindu-Arabic numerals were used extensively in thirteenth-century Italian banking and trading establishments, account-book entries were still kept in roman numerals. Also, documents written with roman numerals carried greater weight in court. And, in 1299, the City Council of Florence made it illegal, at a financial penalty, to keep account books in the new numerals. The rationale behind this resistance was that the new numerals were more subject to fraud than the roman numerals. A Venetian work on bookkeeping, quoted by Menninger, indicates that '. . . the old figures alone are used because they cannot be falsified as easily as those of the new art of computation, of which one can with ease make one out of another, such as turning the zero into a 6 or a 9 . . .' (1969, pp. 426–7).

Another fascinating thing I learned from Menninger is that in the Middle Ages most adults thought of arithmetic as a very advanced, difficult subject. In a lecture delivered in 1517, to students at Germany's University of Wittenberg, the professor appealed:

> Now that I have discussed for you the usefulness of the art of computation, of which there cannot be the slightest doubt, I believe I should also make some brief remarks about the ease with which it can be done. I believe that students allow themselves to be frightened away from this art because of their preconceived notion that it is too difficult . . . those who think they [the elements of computation] are too difficult are greatly in error. This knowledge springs directly from the human mind and appears with full clarity. Therefore its elements cannot be obscure and difficult . . . The rules for multiplication and division, to be sure, require more diligence for their mastery, but their meaning will still be understood very quickly by those who give to them their full attention. These skills, of course, like all others, must be sharpened by practice and experience. (1969, p. 434)

What I didn't get from Menninger was an accurate picture of African numeration systems. The only counting system he identifies is that of the Bushmen, one of the oldest peoples in Africa. He identifies the names of some of their numbers and judges their counting as an 'earlier phase' in the development of number systems. In *Africa Counts*, Claudia Zaslavsky quotes the Viennese anthropologist Marianne Schmidl for a more sophisticated analysis:

> True primitiveness in relation to counting cannot be found. Even where the expression for three or four means 'many', as for example among the Bushmen, it is not true that all counting ends there. They continue counting with the assistance of the words for two and one, and when they do not use higher numbers, they assign a specific meaning to the word 'many' by means of appropriate gestures.
>
> One must be extremely cautious about accepting the accounts of the inability of 'primitive' people to count in higher denominations. ('Primitive' must be taken here, as elsewhere, with a grain of salt!) The absence of counting words by no means indicates a lack of counting concepts, since the number concept and the designations for the numbers need not always coincide. One must consider the level of the economy, the practical need for arithmetic operations, etc. (1973, pp. 13–14)[1]

SOLVED PROBLEMS

Drill

These problems review exactly how, with only ten digits (0, 1, 2, 3, 4, 5, 6, 7, 8, 9), we can write all the numbers. This is because the value of a number depends not only on the value of its digits, but also on the positions, or places, of its digits.

The problems follow the US usage of 'billion' = thousand million; in Britain 'billion' means one million million.

Problem 1 Write each of the following numbers in words:

(Answers below)

(a) 4,302
(b) 252,678
(c) 12,006,024
(d) 15,345,000,460

Solution In order to read and write numbers, you have to be familiar with the names of the place-values. Groups of three place-values each (starting from the ones' place) are separated by commas. In each group, read the number formed by the three digits and then add the name of the group.

hundred trillions	ten trillions	trillions	hundred billions	ten billions	billions	hundred millions	ten millions	millions	hundred thousands	ten thousands	thousands	hundreds	tens	ones
millions			billions			millions			thousands			units		

Answer (a) 4,302 is written 'four thousand, three hundred and two'
(b) 252,678 is written 'two hundred and fifty-two thousand, six hundred and seventy-eight'
(c) 12,006,024 is written 'twelve million, six thousand and twenty-four'
(d) 15,345,000,460 is written 'fifteen billion, three hundred and forty-five million, four hundred and sixty'

Notes
1 Often the comma is omitted in four-digit numbers such as 4302.
2 Technically, the word 'and' is not used in naming whole numbers. When you study decimal fractions you will see that the decimal point is read as 'and'.
3 In many countries, including Haiti, Cape Verde, Cambodia, Vietnam, Guatemala and Greece, the positions of the comma and the period are reversed in writing numbers. So in these countries 'twelve million, six thousand, and twenty-four' would be written 12.006.024; if you are used to this notation, it will initially be confusing to read numbers in the USA and Britain. The key is to understand the *concept* of place-value and learn the values of each place. Then it will be easy to switch to whichever notation is used in the country in which you live.

You try these *(Answers at end of chapter)*

(a) 9,072 (b) 198,467 (c) 425,000,904 (d) 2,000,000,346

Problem 2 Write each of the following numbers in our place-value notation:

(Answers below)

(a) twenty-three thousand, four hundred and sixty-two
(b) two hundred and three thousand and twenty-eight
(c) twenty million and forty-two

Solution Use the place-value chart to make sure that you have written each group of digits in the correct position. For example, in (c), we must use three zeros to get the twenty in the millions place-value group.

Answer (a) 23,462
(b) 203,028
(c) 20,000,042

You try these *(Answers at end of chapter)*

(a) forty-eight thousand, nine hundred and seventy-five
(b) four hundred and six thousand and fifty-two
(c) forty million and seventy-nine

Problem 3 Write in expanded notation: *(Answers below)*

(a) 342
(b) 4,674
(c) 3,098

Solution Expanded notation means writing the number to illustrate the place-value of its digits. The purpose of this exercise is to emphasize what it means to say that the value of a number depends on the place-values of its digits.

Answer

(a) $342 = 300 + 40 + 2$
$= 3 \text{ hundreds} + 4 \text{ tens} + 2 \text{ ones}$
$= (3 \times 100) + (4 \times 10) + (2 \times 1)$

(b) $4{,}674 = 4{,}000 + 600 + 70 + 4$
$= 4 \text{ thousands} + 6 \text{ hundreds} + 7 \text{ tens} + 4 \text{ ones}$
$= (4 \times 1{,}000) + (6 \times 100) + (7 \times 10) + (4 \times 1)$

(c) $3{,}098 = 3{,}000 + 0 + 90 + 8$
$= 3 \text{ thousands} + 0 \text{ hundreds} + 9 \text{ tens} + 8 \text{ ones}$
$= (3 \times 1{,}000) + (0 \times 100) + (9 \times 10) + (8 \times 1)$

Notes

1 Problem (c) illustrates the importance of the digit zero. Zero ensures that the 3 in 3,098 represents 3 thousands. Without zero, how could we distinguish 3,098 from 398? We write (0×100), which equals zero, in the expanded notation, to emphasize that there are no hundreds in 3,098.

2 The key to expanded notation is to make the abstract place-values concrete. Each of the three forms of expansion (that is, $42 = 40 + 2 = 4 \text{ tens} + 2 \text{ ones} = (4 \times 10) + (2 \times 1)$) does this, so you should use whichever one is clearest to you.

You try these (*Answers at end of chapter*)

(a) 423 (b) 5,858 (c) 203

Thought Problem 4 What is the relationship among the place-values? *(Answer below)*

Solution Each place is ten times larger than the place to its immediate right. For example, $1{,}000 = 10 \times 100$ and $10{,}000 = 10 \times 1{,}000$. This is why our place-value system is called a *base-10* or *decimal system* (the prefix *deci* means ten).

When we count objects, we group them in tens or in multiples of ten (for example, hundreds, thousands). Having 492 apples means you have 4 groups of one hundred apples each, 9 groups of ten apples each, and 2 loose apples. If you have 10 groups of any place-value, you have 1 group of the next larger place-value. For example, 10 groups of one hundred apples each is the same as 1 group of one thousand apples.

You try this *(Answer at end of chapter)*

Write the decimal number which represents 10 groups of 100,000 apples each.

Applications

Problem 5 Write the number in the following sentence in words: 'According to the Children's Defense Fund, there were 820,099 handicapped[2] children in the USA not served by public school programmes in spring 1976.' *(Answer below)*

Solution I believe numbers lose some of their impact when we skip over them in reading. In Volume Two we will focus on getting a concrete idea of what gigantic numbers mean. Here, we will begin that process by practising reading numbers in context.

Answer There were eight hundred and twenty thousand and ninety-nine handicapped children not served by public school programmes in spring 1976.

You try this *(Answer at end of chapter)*

There were 1,070,312 children (11–14 years of age) arrested in 1977.

Note: For those students interested in more information about topics covered in the real-life applications, this book concludes with a list of relevant organizations in the USA and Britain.

UNSOLVED PROBLEMS

(Answers at end of chapter)

Drill

1 Write each of the following numbers in words:

(a) 4,672 (b) 4,070
(c) 4,000,050 (d) 4,524,007,060.

2 Write each of the following numbers in our decimal place-value notation:

(a) eight thousand, four hundred and seventy-five

(b) eight thousand and seventy
(c) forty-nine million, four hundred and twenty-five
(d) two hundred and three billion and seventy thousand

3 Write in expanded notation:

(a) 2,070
(b) 124,004

Thought

4(a) Describe the 'number puns' or 'plays on numbers' in the title of the following article (*In These Times*, 18 March 1981).

(b) How are these 'puns' related to the concept of place-value?

'925: 20 million 2 go

'Working Women, a national network acting on behalf of 10,000 women office workers, joined forces this month with the 650,000-member Service Employees International Union to form a new clerical organizing group – District 925, a play on "9 to 5". The new union district has a potential membership of nearly 20 million workers, 90 percent of whom are unorganized. Working Women will also continue as an independent organization.'

5(a) What is the joke in the following political cartoon? (New York *Guardian*, 8 June 1983)

(b) What place-value concepts do you have to understand in order to understand this joke?

6 This is an old trick which was played on me years ago (and which I fell for!) Quickly read someone the following list of numbers, asking them to say the next higher number after each number you recite:

66; 12; 123; 8; 129; 415; 613; 1,175; 58; 4,099.

Most people give 5,000 as the next number after 4,099.

(a) What is the next number after 4,099?

(b) Why do you think most people say 5,000?

7 According to Karl Menninger (p. 403), throughout history poets and philosophers have been fascinated by zero, the magical digit which creates so much and yet stands for nothing. He quotes a Hindu saying: 'Ten men live in such a way that they allow one to take precedence. Without this one they have as little meaning as zeros unless they are preceded by a "one". Write briefly about what this saying means:

(a) literally, that is, in terms of the use of zero in the place-value system;

(b) figuratively, that is, how the literal fact about the use of zero in the place-value system relates to people organizing themselves in a group;

(c) do you agree or disagree with this saying, and why?

8 Read the following book review by John Allen Paulos of *From One to Zero: A Universal History of Numbers* (*New York Times*, 29 September 1985). Write about:

(a) which information was most interesting to you and why;

(b) which information was confusing, and what questions you need answered before you can understand that information.

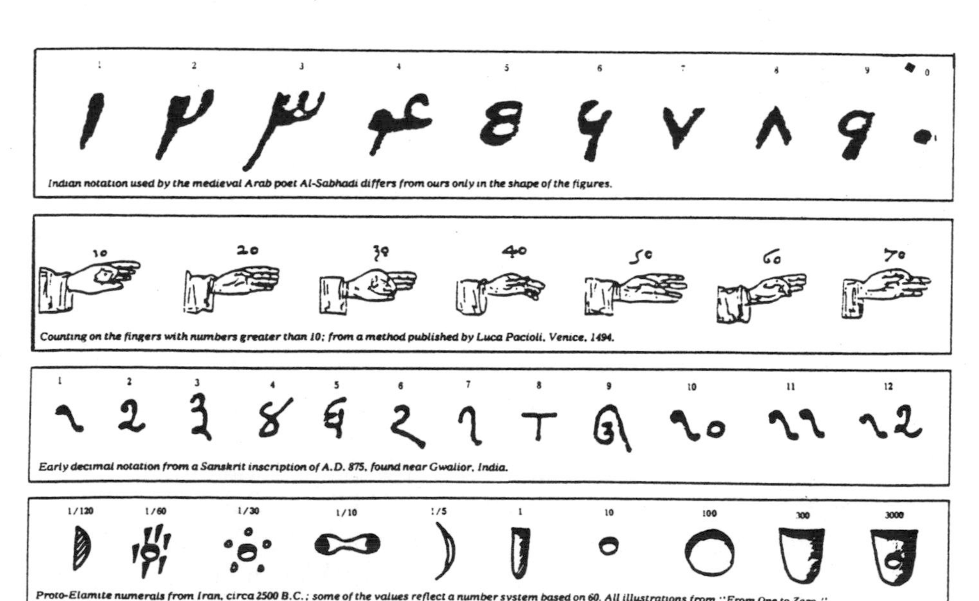

Indian notation used by the medieval Arab poet Al-Sabhadi differs from ours only in the shape of the figures.

Counting on the fingers with numbers greater than 10; from a method published by Luca Pacioli, Venice, 1494.

Early decimal notation from a Sanskrit inscription of A.D. 875, found near Gwalior, India.

Proto-Elamite numerals from Iran, circa 2500 B.C.; some of the values reflect a number system based on 60. All illustrations from "From One to Zero."

'A German merchant of the fifteenth century asked an eminent professor where to send his son for a good business education. The professor responded that German universities would be sufficient to teach the boy addition and subtraction but that he would have to go to Italy to learn multiplication and division. Before you smile indulgently, try adding the roman numerals CCLXVI, MDCCCVII, DCL and MLXXX without first translating them.

'The Moroccan-born French mathematics professor Georges Ifrah relates this anecdote in *From One to Zero: A Universal History of Numbers*, which has been glowingly reviewed in France. The book is an exhaustive and at times exhausting history of numerals (number symbols) and numeration systems from prehistoric times to the Renaissance, when the Hindu-Arabic system we use today came to worldwide dominance. The heroes of his tale are the nameless scribes, accountants, priests and astronomers who discovered the principles of representing numbers systematically.

'These principles – abstract symbolization (as opposed to concrete representations with pebbles, say), a multiplicative base for the system (the numeral 134 in a base 12 system, for example, would represent the number we ordinarily write as 184: [1 × 12 × 12] + [3 × 12] + [4 × 1]), positional notation (826 is very different from 628 or 682), and the holy grail, zero (allowing one to distinguish easily between 36, 306, 360 and 3,006) – are an essential though almost invisible part of our cultural heritage. It is the great virtue of Mr Ifrah's book that he makes this heritage not only visible but vivid.

'He describes various concrete means of indicating numbers – pebbles of different sizes, knotted skeins of colored threads, the ubiquitous abacus and counting board and finally the most personal of personal computers: human hands (and feet). Counting on our fingers or on our fingers and toes is an almost universal phenomenon that ultimately gave rise to the most commonly written bases. Our base 10 system is certainly a consequence of this, while the French words for twenty, eighty and ninety – *vingt, quatre-vingt* and *quatre-vingt-dix* – suggest an older base 20 system. The Mayas, one of the four peoples to invent the principle of positional notation, also used a base 20 system 1,500 years ago to create calendars more accurate than the Gregorian one we use today. Even the ancient Babylonian-Sumerian base 60 system, which survives in our measurement of time, angles and geographical position, was probably derived from finger counting.

'Mr Ifrah provides an interesting chapter on numerology, the assigning of numerical equivalents to letters and the consequent reading of hidden meanings into words and phrases. A Roman Catholic mystic, for example, noted that Martin Luther had to be the

Antichrist because his name had the dreaded value 666, to which Luther's disciples replied that this was the value of the inscription on the papal crown.

'There are also digressions on alphabets and hieroglyphs, history and archeology, and over 350 illustrations by the author himself. Many of the sidelights are fascinating: Pascal's triangle was known to the Arabs and Chinese more than three centuries before Pascal. Wooden sticks with matching notches (one for the creditor and one for the debtor) were long used as "credit cards" in many parts of the world. But too much of this long book reads like a collection of appendices, and I often found myself saying "enough already" as Mr Ifrah piled up his historical documentation.

'To resume the main thread of the story, the Chinese about 2,000 years ago and the people of south India about 1,500 years ago independently invented a written positional numeration system based on powers of 10. Shortly thereafter, the Indians went farther and invented a zero that transformed the art of representing and manipulating numbers forever (the Mayas and Babylonians had a sort of nonfunctional zero). The Chinese borrowed the notion from the Indians, as did the Arabs, who eventually communicated the system to western Europe. Mr Ifrah writes that the invention of this Hindu-Arabic number system was one of the most important technical discoveries of mankind, ranking with the invention of the wheel, fire and agriculture, and he is probably right.

'By the way, the numbers mentioned in the first paragraph add up to MMMDCCCIII.'

9(a) Create and solve a three-question quiz that reviews the major concepts learned in this chapter.

(b) Write briefly about which of your questions is the most difficult and why; which question is easiest and why.

Applications

Write the numbers in each of the following sentences in words.

10(a) According to *WIN Magazine* (1 January 1981), during the July 1980 draft registration period for men born in 1960 and 1961, large numbers of men refused to sign up. Significantly, the resistance was not exclusively, or even primarily, from the larger cities of the east and west coasts. For example, out of 25,500 required to register in Greater Kansas City, only 18,630 registered. For another example, out of 28,500 required to register in the Rockford–Springfield–Peoria, Illinois area, only 18,140 registered.

(b) Rio-Tinto Zinc (RTZ) is a British-based mining conglomerate with operations in more than 40 countries, including Namibia, South

Africa and Australia. In 1986, RTZ's turnover was £4,343 million, with profits of £245 million. It is the world's second largest mining company, employing 82,551 people. Among its more notorious activities is the illegal mining of uranium at Rossing in Namibia. This mining continues through 1988 in contravention of United Nations Resolution 435, which forbids the extraction of Namibia's resources while South Africa continues to occupy the country. The lack of trade-union rights, poor safety and outright discrimination are fairly typical practices of transnational corporations.

11 Read the following Ann Landers column (*Boston Globe*, 17 May 1982). Write briefly about:

(a) the main point she is making;

(b) how she uses numbers to support her main point;

(c) which of her number arguments you think is most powerful and why;

(d) what kinds of numerical data would someone who disagrees with her main point want to collect. In addition, you could research any numerical data currently available about the nuclear threat.

'Dear Ann Landers:

'Can you stand another letter about the woman who didn't know whether or not to leave her gold crowns to relatives when she dies, for fear she might need them when she returns to life for the resurrection?

'If the two strongest nations in the world, the United States and Russia, don't agree to put a freeze on nuclear weapons she won't have to worry about her teeth or anything else because they will be vaporized in a matter of minutes – along with millions of people.

'Talk about a limited nuclear war and plans for civilian defense are insane. It would be impossible to evacuate the cities. Where would people go? What would they eat? What would they drink? Who would take care of them? The physicians and hospitals would be blown to smithereens.

'People believe in you. For the love of mankind and its survival, please address yourself to this issue.

'– Terrified In DC

'Dr James E. Muller of the Harvard Medical School said, "The horror of nuclear war is so great that many people choose to deny it exists.

'"An all-out attack on the United States could kill as many as 150 million Americans. Their immune systems, weakened by radiation, would succumb to fatal diseases. At least 80 percent of the doctors would be incinerated."

'As the Rev. Theodore Hesburgh said at a UCLA peace rally,

"The living would envy the dead."

'The *New Republic* pointed out in a recent editorial that the global arms budget for all countries is now $550 billion a year. About $100 billion is earmarked for nuclear weapons. When one considers that $500 billion equals the entire annual income of the poorer half of the earth's four billion people, one begins to grasp the magnitude of that expenditure. Meanwhile, we are cutting programs that benefit the elderly, the handicapped and the poor. More cheery news: our Secretary of the Treasury Donald Regan says, "Our economy is dead in the water."

'The standard reference is the Hiroshima bomb. It destroyed the city with the equivalent of 12,500 tons of TNT. (This is 12.5 kilotons.) The newest nuclear bomb is not measured in kilotons, but in megatons. The yield would be equivalent to approximately 12 million one-ton trucks filled with TNT.

'The Poseidon submarine carries 16 missiles, each with 10 warheads. Each warhead has three times the explosive force of that single bomb dropped on Hiroshima. We can already kill every Russian seven times. Now we are trying to build our arms supply so we can kill each Russian 14 times. The United States and other major powers are spending themselves broke on a war we dare not let happen.

'Jonathan Schnell wrote in the *New Yorker*, "The machinery of destruction is in place, poised on a hair-trigger, waiting for the button to be pushed by some misguided or deranged human, or for some faulty computer chip to send out the instructions to fire."

'I implore every person to sign his or her name across this column and mail it to President Ronald Reagan, the White House, Washington, DC. An overwhelming response might prevent a nuclear holocaust that would mean the end of all life on this planet. Do it TODAY. Nothing on your calendar can be more important.'

12 Look through a newspaper or magazine for various uses of numbers and write each of the numbers you find in words.

Recreation

13 According to Morris Kline, the sixth-century BC Greek mathematician Pythagoras's followers

> 'considered the number "one" as the essence or very nature of reason, for reason could produce only one consistent body of doctrines. The number "two" was identified with opinion, clearly because the very meaning of opinion implies the possibility of an opposing opinion, and thus of at least two. "Four" was identified with justice because it is the first number aside from the trivial case of 1 which is the product of equals.' (1972, p. 59)

In a more sensual view, a contemporary mathematician states, 'I picture 7 as dark and full of liquid, like oil when it oozes from the ground. Three is lumpy and hard, but dark also, and 4 is soft and doughy and pale. Five is pale, but round like a ball, and 6 is like 4 only richer, like cake instead of dough' (Friedberg, 1968, p. 11).

Write a creative essay, poem, or short story about what the digits mean to you.

Read the following book review by Christina Robb of George Weinberg's *Numberland: A Fable* (*Boston Globe*, 27 February 1987).

(a) Why do you think the numbers think they are immortal?
(b) Why can't the rescue troops find the last number?
(c) Why do you think the 6's search for the truth about people, numbers and death nearly tears Numberland apart?
(d) What truth do you think he learned that finally brought Numberland back together as a sadder, wiser, much dreamier place?
(e) Write your own creative essay, poem or short story about an adventure in Numberland.

A fable made up of numbers

A banking friend once took Citibank's six-month executive training course. She and a dozen other young bankers were closeted in a warehouse on Long Island for 12-hour days and then sent home with numbers to punch, grind, analyze and otherwise fixate on till dawn. At the end of six months, she could do an arithmetical operation in her head faster than a calculator. She was also dreaming about numbers during the few hours of sleep she got at night. She wondered a little sadly whether she had become so computer-like that her numerical dreams made her ineligible for psychoanalysis and disproved Freud's theory about dreams as wish-fulfillments.

'But Freud left ample room in his dream theory for the residues of the day's activities. All he would have said of her numerical dreams was that she was using numbers to bear the symbolic freight of more usual dream images, I told her. If she let her mind drift, she would find she could make Freudian free-associations to the numbers in her dreams. What did the 88 she had dreamed about the night before remind her of, for instance? She smiled and admitted that Freud was right.

'George Weinberg is a psychotherapist who has written about psychology. He is also co-author of a statistics text, and he has had such an intensive relationship with numbers that he has dreamed up a whole country and civilization of numbers in broad daylight.

'Numberland is a world where little numbers get along with big ones in an almost perfectly orderly way. Its republican seat of government is Numeropolis, where the smaller numbers live and represent the ever-larger figures who occupy the country's ever-more-hinterlands.

'A General, 777* (the asterisk is to remind constituents of his military rank), is governing Numberland when the fable begins, but he is getting ready to face a professor, 1,000, in the imminent general election. Things look good for the incumbent defender of law and order until an inquisitive little digit called 6 starts asking about people and death.

'Part of the reason the numbers in Numberland remain so orderly is because they think they are immortal. Though a few of them have heard of people, another truism of Numberland tells them that people and numbers can never meet. One very respectable school of thought holds that people don't exist at all. Scholars alone bother with us – until 6 gets curious.

'Six discovers from his scholarly friend 44, who is chairman of the populicity department at the university, that people and numbers occupy the same geography, but it is held that they can't communicate because people are mortal and numbers are immortal. Numbers have revealed themselves to only two or three geniuses in human history, but geniuses die and all numbers remain alive. One living genius is 9-year-old Elizabeth Smith, who is about to give up studying numbers because other people think she is crazy for doing it.

'In a dream that night (or some dreamlike out-of-body experience), 6 is translated from Numberland to Elizabeth's room. He meets her, talks to her, convinces her to keep studying numbers and is wafted back to Numberland a changed digit. He is convinced now that numbers and people can meet, and he has a nagging doubt about the immortality of numbers.

'As news of his experience gets around, his doubts about the deathlessness of numbers spread, too. In response to this creeping apprehensiveness, the General promises a parade of all the numbers from first to last, to prove they are all present and accounted for forever. But his crack rescue troops can't find the last number. The wily Professor seizes this moment to present his own, winning campaign promises: to find the biggest number and educate all numbers to a certainty of their immortality. But 6 is still asking questions, and getting disturbing answers.

'Little 6's search for the truth about people, numbers and death nearly tears Numberland apart. But the truth he finally learns brings it back together as a sadder, wiser, much dreamier place.

'Weinberg has told a clever and poignant fable that will humanize the hardest-headed number-cruncher and numerize the most numerophobic non-genius.'

Projects

15 Read the science-fiction short story 'The masters' in *The Wind's Twelve Quarters* by Ursula K. LeGuin (Bantam Books, 1975). Write a brief review of the story, indicating how its theme is related to this section.

16 Painters have also used their creativity to represent numbers. The contemporary artist Jasper Johns, in particular, often uses numbers as a theme. Study his number works and prepare either a report or a slide show on them, or create some of your own number art.

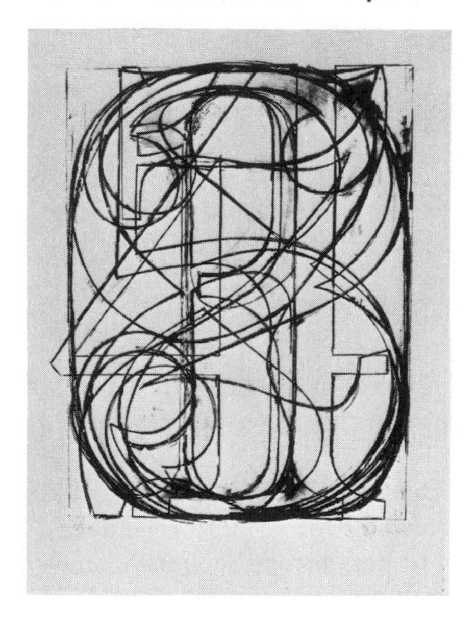

Jasper Johns, 0 Through 9, 1960. Collection, The Museum of Modern Art, New York.

***Jasper Johns*, 0 Through 9**

REVIEW QUIZ (the meaning of whole numbers)

(*Answers at end of chapter*)

1 Write in words: 6,005,402.

2 Write in decimal place-value notation: forty million, three hundred and fifty.

3 Write in expanded notation: 42,007.

4 Write the numbers in each of the following sentences in words:

(a) Out of 47,000 men required to register for the draft in the Greater Boston area, only 32,950 registered.

(b) From the Ann Landers column:
'An all-out nuclear attack on the USA could kill as many as 150,000,000 North Americans.'
In the early 1980s 'the global arms budget for all countries is $550,000,000,000 a year.'

ANSWERS: CHAPTER 9

Solved problems ('you try these')

1(a) nine thousand and seventy-two
(b) one hundred and ninety-eight thousand, four hundred and sixty-seven
(c) four hundred and twenty-five million, nine hundred and four
(d) two billion, three hundred and forty-six (in USA); or two thousand million, three hundred and forty-six (in UK)

2(a) 48,975
(b) 406,052
(c) 40,000,079

3(a) $(4 \times 100) + (2 \times 10) + (3 \times 1)$
(b) $(5 \times 1{,}000) + (8 \times 100) + (5 \times 10) + (8 \times 1)$
(c) $(2 \times 100) + (0 \times 10) + (3 \times 1)$

4 $10 \times 100{,}000 = 1{,}000{,}000$. Millions is the place-value to the immediate left of hundred thousands, so one million is 10 times one hundred thousand.

5 one million, seventy thousand, three hundred and twelve (Note that 1977, the year, is read 'nineteen seventy-seven' or 'nineteen hundred and seventy-seven'.)

Unsolved problems

1(a) four thousand, six hundred and seventy-two
(b) four thousand and seventy
(c) four million and fifty
(d) four billion, five hundred and twenty-four million, seven thousand and sixty

2(a) 8,475
(b) 8,070
(c) 49,000,425
(d) 203,000,070,000

3(a) $(2 \times 1{,}000) + (0 \times 100) + (7 \times 10) + (0 \times 1)$
(b) $(1 \times 100{,}000) + (2 \times 10{,}000) + (4 \times 1{,}000) + (0 \times 100) + (0 \times 10) + (4 \times 1)$

4(a) The name of the union, from the average working hours of its members, is '9 to 5', so because 'to' is pronounced the same as 'two', the union chose the number 925. The title of the article also reverses this pun by using the number '2' instead of the word 'to' in '20 million to go'.
(b) To get the pun we read '925' as 'nine two five', but in place-value we read this number as nine hundred and twenty-five. So if the place-values were considered the pun would disappear.

5(a) The US Army keeps increasing the number of advisors they say are in El Salvador, and the cartoon suggests this is done illegally, without consent of Congress, but just by tampering with the numbers.
(b) The tampering is done by adding zeros, pushing the place-values of the fives higher and higher, first to five hundred and fifty (550), then five thousand, five hundred (5,500), then fifty-five thousand (55,000), and lastly to five hundred and fifty thousand (550,000).

6(a) 4,100
(b) I said 5,000 because when I was listening and heard 99, and didn't visually see the zero in the hundreds' place, I treated 4,099 as if it were 4,999, where the next number would be 5,000.

7(a) When '1' precedes a 'zero', we get 10 (ten).
(b) The saying implies that without a leader, a group has no meaning since zeros without a 1 in front just add up to zero (00000 etc. = 0).

10(a) For example, out of twenty-five thousand and five hundred required to register in Greater Kansas City, only eighteen thousand, six hundred and thirty registered. Out of twenty-eight thousand and five hundred required to register in the Rockford–Springfield–Peoria area, only eighteen thousand, one hundred and forty registered. Notice that in the USA and Britain numbers for dates are written in different ways, for example January 1, 1981 (in the USA) and 1 January 1981 (in Britain). So 3/4/88 can mean March 4, 1988 (in the USA) or 3 April 1988 (in Britain).

(b) Forty countries; four thousand three hundred and forty-three million pounds or, more commonly now in English, four billion, three hundred and forty-three million pounds (we'll learn why in Chapter 11); two hundred and forty-five million pounds; eighty-two thousand five hundred and fifty-one people; Resolution four hundred and thirty-five. Notice that we usually read '1986' as 'nineteen eighty-six', not as 'one thousand nine hundred and eighty-six'.

Review quiz

1 six million, five thousand, four hundred and two

2 40,000,350

3 $(4 \times 10{,}000) + (2 \times 1{,}000) + (0 \times 100) + (0 \times 10) + (7 \times 1)$

4(a) Out of forty-seven thousand men required to register for the draft in the Greater Boston area only thirty-two thousand, nine hundred and fifty registered.

(b) An all-out nuclear attack on the USA could kill as many as one hundred and fifty million North Americans.

The global arms budget for all countries is five hundred and fifty billion dollars a year.

10 The meaning of fractions

If fractions have been confusing to you in the past, it may be because people hold a common misconception that fractions have only one meaning. On the contrary, fractions are used to represent a number of different concepts.

1 FRACTIONS ARE USED TO REPRESENT PARTS OF A WHOLE

(a) Pizza pies, for example, are frequently divided into 8 equal slices. If you order 3 slices, the fraction $\frac{3}{8}$ represents the portion of the entire pizza pie that you ordered. '$\frac{3}{8}$' is read 'three eighths'; the bottom number (the *denominator*), which in this case is 8, represents the number of equal pieces into which the whole is divided; the top number (the *numerator*), which in this case is 3, represents the number of equal pieces to which you are referring.

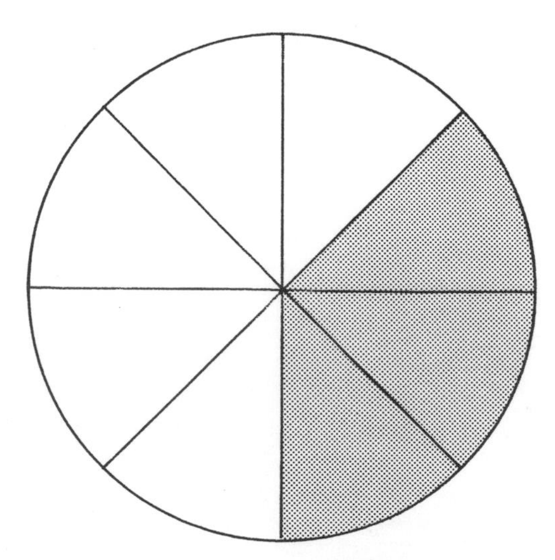

(b) A ruler, for another example, is sometimes divided into inches where each inch is further broken down into 4 equal parts. Each of the parts in the picture is $\frac{1}{4}$ of an inch, 1 of 4 equal parts into which the inch is divided. '$\frac{1}{4}$' is read 'one fourth' or 'one quarter'; the nail in this case measures $1\frac{1}{4}$ in (read 'one and a quarter inches'). By breaking up each inch into fractional parts we can more accurately represent lengths that are between the inch divisions. Inches on rulers are also commonly broken up into 8 equal parts (eighths) and 16 equal parts (sixteenths).

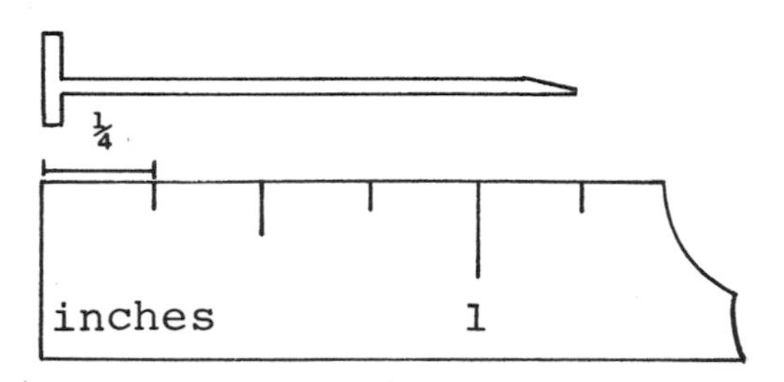

2 FRACTIONS ARE USED TO REPRESENT COMPARISONS BETWEEN GROUPS

We can express the fact that in 1977, 18 out of the 435 members of the US House of Representatives were women, as the fraction $\frac{18}{435}$. We say that the *ratio* of women representatives to the total number of representatives was 18 to 435, or that 18 out of 435 representatives were women, or that eighteen 435ths of the representatives were women.

3 FRACTIONS ARE USED TO REPRESENT DIVISION

Suppose you wanted to share equally 4 pizza pies among 3 people. To find out how many pies each person will get, we divide 4 by 3. As you will learn in Volume Two, this division can be represented as $3\overline{)4}$ or $4 \div 3$ or $\frac{4}{3}$. As you will learn in this volume (Chapter 14), the fraction $\frac{4}{3} = 1\frac{1}{3}$. In this case, each person would get one pizza pie and one third of a pizza pie, or $\frac{4}{3}$ pizzas.

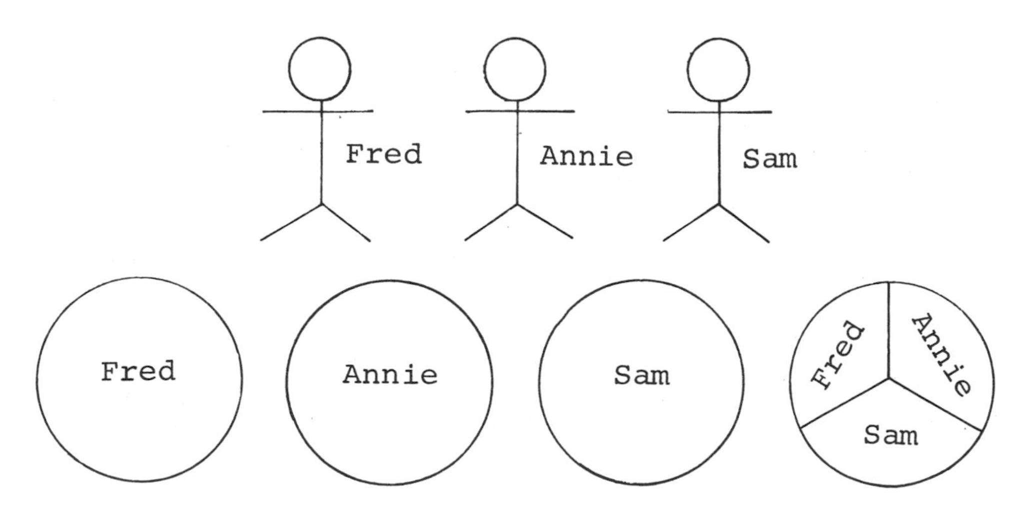

SOLVED PROBLEMS

Drill

Problem 1 Name the fraction which represents the circled part of the following diagram. *(Answer below)*

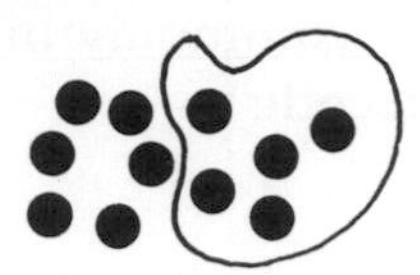

Solution In this case the entire group of 11 dots represents a whole containing 11 equal parts. Therefore, 11 is the denominator. Since 5 dots are circled, or 'referred to', 5 is the numerator. 5 out of 11 dots are circled.

Answer $\frac{5}{11}$

You try this *(Answer at end of chapter)*

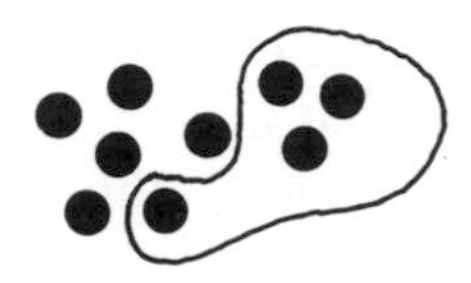

Problem 2 Shade in $\frac{2}{3}$ of the following diagram. *(Answer below)*

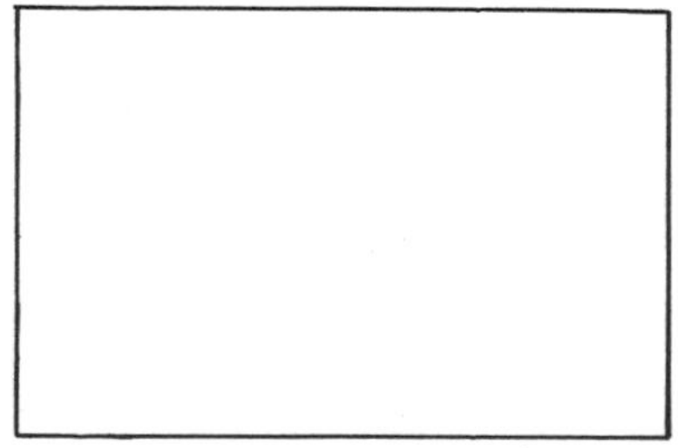

Solution The denominator 3 indicates that the whole should be divided into 3 equal parts. This can be done correctly in a number of different ways. For example,

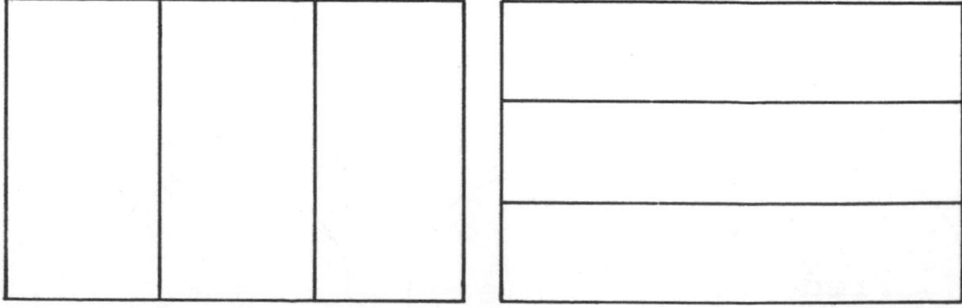

The numerator indicates that 2 of these parts should be shaded. You can shade any 2 of the parts.

Answer Two possible answers are:

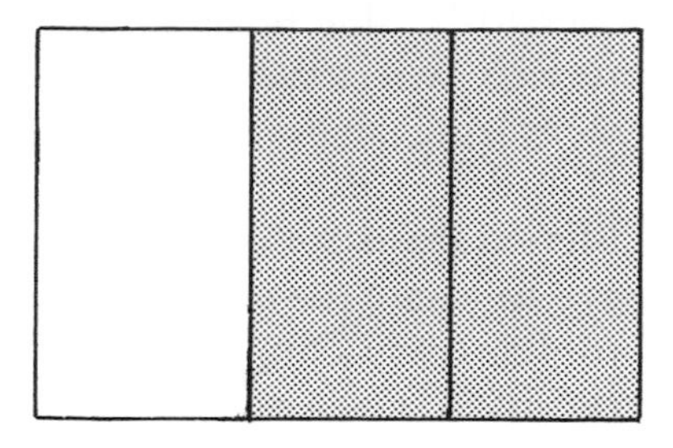

or:

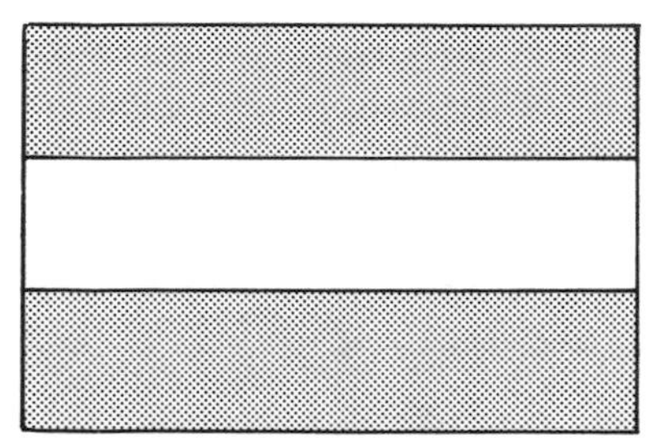

You try this *(Answer at end of chapter)*

Shade in $\frac{3}{5}$ of the following diagram. *(Answer at end of chapter)*

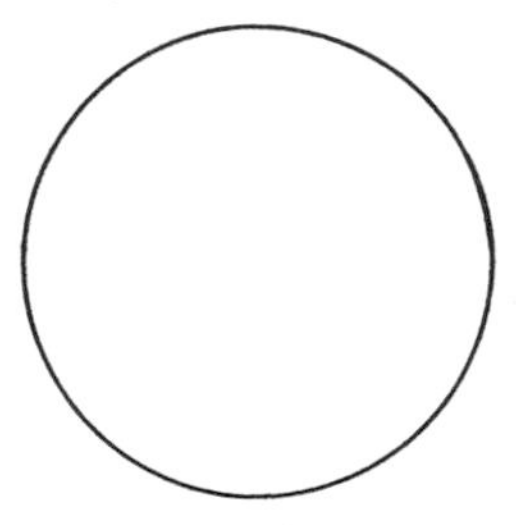

Thought Problem 3 Find the error pattern (what is correct, what is incorrect) in the following three solved problems. Then redo the problems correctly. Shade in $\frac{1}{2}$ of each of the following diagrams *(Answer below)*:

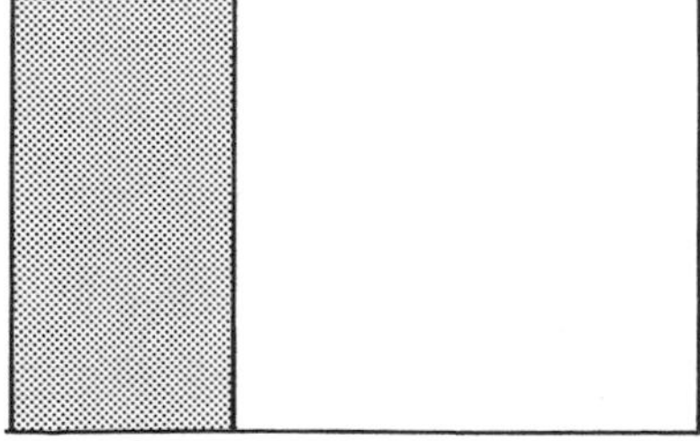

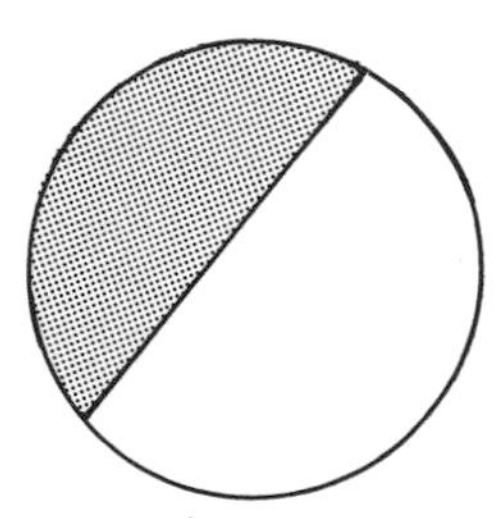

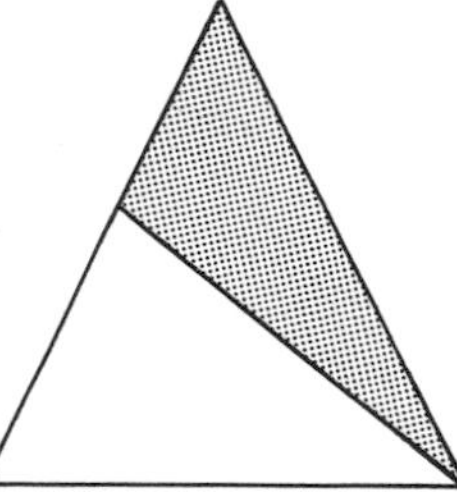

Solution All wrong answers (except those guessed wildly from pure anxiety) involve some correct, logical reasoning. Analysing error patterns gives you thoughtful reinforcement of maths skills because you have to understand the rules thoroughly in order to be able to determine what was done correctly and precisely where the solver went wrong. You also get practice recognizing many correct methods of solving problems, because the solver may be using an approach which is totally different from yours. You cannot just say 'it's *all* wrong', but have to locate the exact error. I hope that these error pattern problems will emphasize to you that you are not 'stupid' if you make an error, and will help you develop respect for your own thought process as you solve problems.

Answer In this case, the person realized that the denominator means divide the whole into 2 parts and that the numerator means shade one of those parts. The mistake was that the whole cannot be divided arbitrarily into 2 parts, but must be divided into 2 *equal* parts. One correct way to solve the problem is:

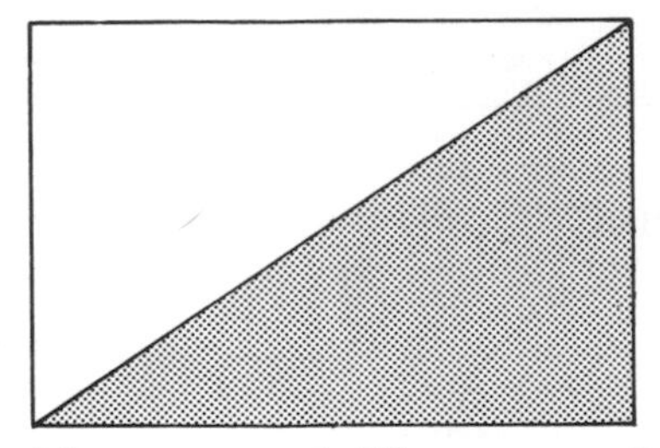
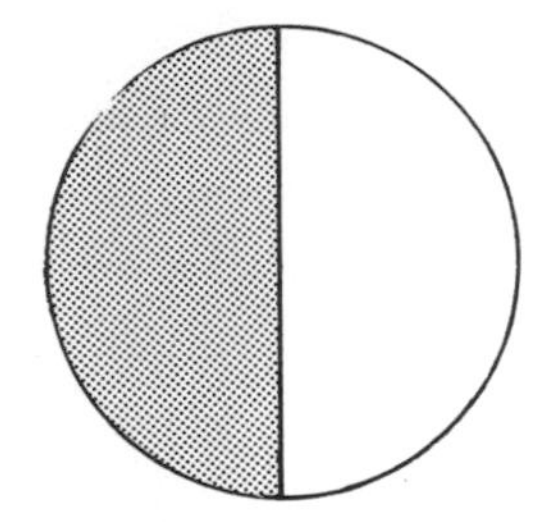
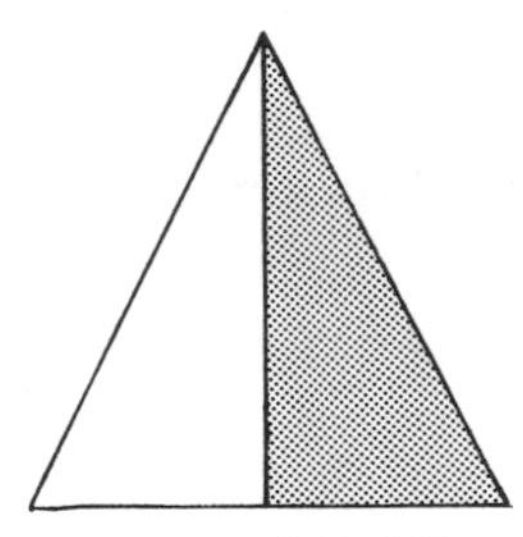

Notes 1 There are, of course, many other correct ways of dividing a whole into halves. For example, a clever division of a circle into 2 equal parts is the Chinese yin-yang symbol, representing equal, but opposite, balanced forces.

2 You can carry these error pattern problems further by discussing how to convince the solver that parts of the method s/he used were wrong, and how to teach the solver the correct method that fits in best with his/her understanding. In this case, you could start by asking the solver what the concept of $\frac{1}{2}$ means to him/her and gradually lead up to the analysis in the above answer. Then you could ask the solver to redo the problems correctly, maybe trying to do each diagram in a few different correct ways. Finally,

you could create a different, related problem, such as 'shade in $\frac{1}{3}$ of each of the following diagrams', which would check the solver's general understanding of the fact that the denominator means 'divide the whole into that many *equal* parts'.

You try this *(Answers at end of chapter)*

Shade in $\frac{3}{4}$ of each of the following diagrams:

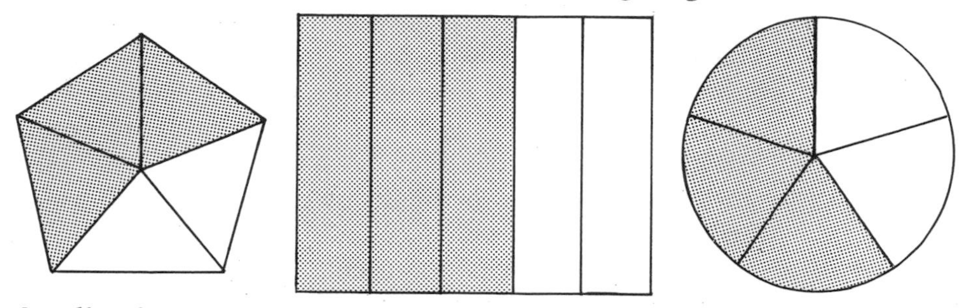

Applications

Problem 4 Represent the length of the following nail using fractions. *(Answer below)*

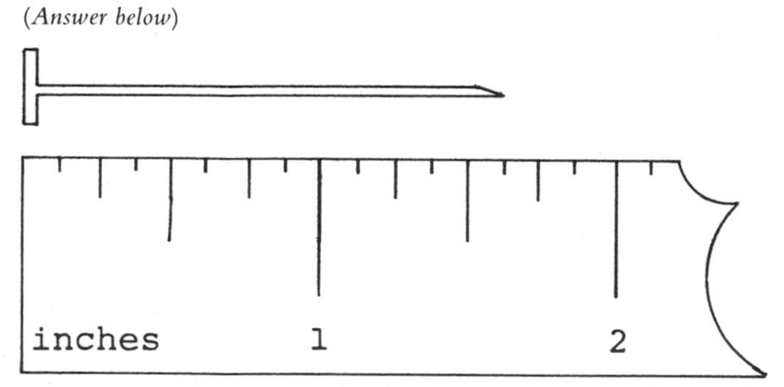

Solution Each inch on this ruler is broken up into 8 equal parts. Therefore, each small division represents $\frac{1}{8}$ in.

Answer The nail is $1\frac{5}{8}$ inches long.

Note The lines indicating inches on a ruler are usually the longest and the inches are numbered. The next-to-longest lines break each inch into 2 equal parts, or into half inches.

The next shorter lines break the halves in half, or break the whole inch into quarters.

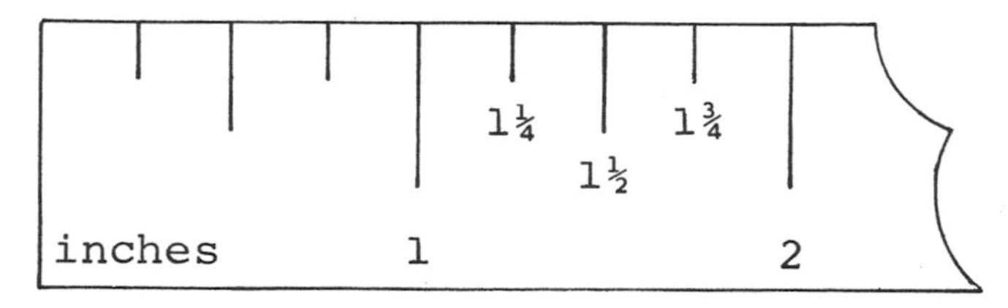

If, as in this problem, a ruler is divided into eighths of an inch, still shorter lines break each quarter inch in half, breaking the whole inch into 8 equal parts.

You try this *(Answer at end of chapter)*

Problem 5 In 1977, the US Senate was composed of 96 white senators, 1 black senator, and 3 senators from other racial groups. What fraction of all the senators were people of colour?

Solution This question asks you to compare the number of senators of colour to the total number of senators. Another way of phrasing this question is 'What part of the entire group of senators are people of colour? A helpful hint for expressing ratios is first to write the English words in fraction form, and then fill in the numbers.

The number of senators of colour compared to the total number of senators forms the fraction:

Answer $\frac{\text{senators of colour}}{\text{total senators}} = \frac{4}{100}$

Notes

1 In English, we would say that in 1977, 4 out of the 100 senators were people of colour. We might also compare this to the fact that in the US population about 14 out of every 100 people were people of colour, so people of colour are not represented proportionally to their numbers in the population.

2 We had to use addition to find the total number of senators of colour (1 + 3). To find the total number of senators, we could either have added (96 + 1 + 3), or have known that each of the 50 states elects 2 senators, for a total of 100 senators.

3 The key difficulty with ratio problems is to know which number is the numerator and which number is the denominator. Whenever you are comparing a group of one kind to the entire group, as in this case, the part (senators of colour) is the numerator, and the entire group (total senators) is the denominator. Whenever you are asked for a ratio of one quantity *compared to* another quantity, the first quantity is the numerator. For example, if you were asked for the ratio of white senators to (or compared to) black senators, your fraction would be set up:

$$\frac{\text{white senators}}{\text{black senators}} = \frac{1}{96}$$

and you would say the ratio of white to black senators is 96 to 1. With experience, you will gain confidence in the translations of various English 'fraction sentences' into math.

4 We sometimes want to put numerical data into fraction form because fractions give a powerful picture of the relationship between two groups. A pie sliced into 100 equal pieces, with only 4 shaded, gives a dramatic picture of how small a portion of the senators were people of colour.

You try this *(Answer at end of chapter)*

What fraction of all the senators were black?

Problem 6 Write the fraction that represents the amount of money each person gets when $12 is being equally shared among 5 people. *(Answer below)*

Solution $12 is being divided into 5 equal pieces, so the division problem is $12 \div 5$.

Answer $\frac{12}{5}$ dollars

Note At present you may only have an intuitive understanding of these fraction division problems because we will not study the meaning of division until Volume Two. Learning is never totally linear; you cannot completely understand every aspect of one topic before moving to the next. Almost always, you have to accept some ambiguity the first time through any new topic. As you learn more and more maths, your understanding of all the math you have learned will become clearer and clearer.

You try this *(Answer at end of chapter)*

Write the fraction that represents the amount of pizza each person gets when 8 pizzas are being shared equally among 7 people.

UNSOLVED PROBLEMS

(Answers at end of chapter)

Drill

1 In your own words, summarize the various meanings of fractions. Include your own example of each meaning.

2 Name the fraction which represents the circled or shaded part of each diagram.

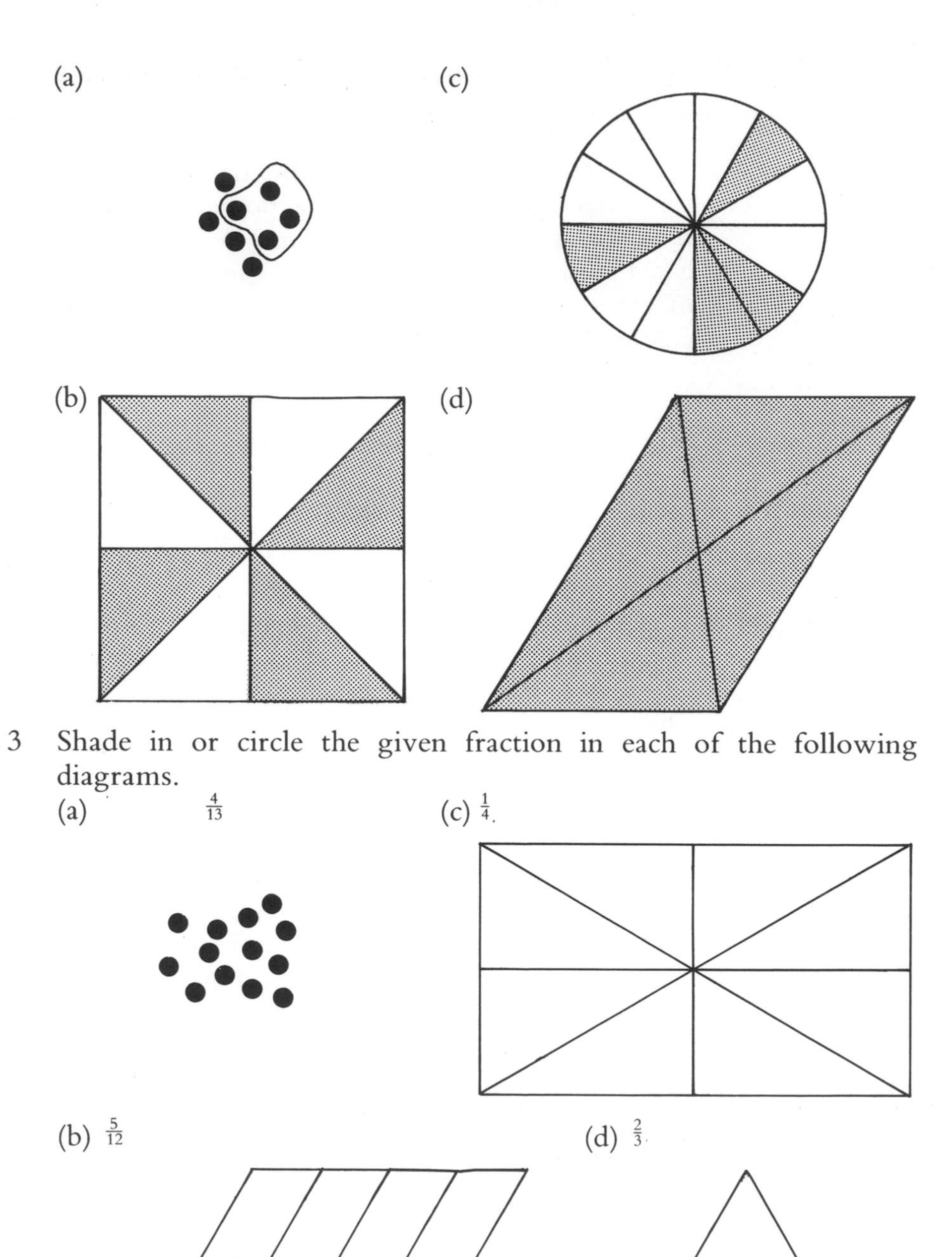

3 Shade in or circle the given fraction in each of the following diagrams.

Thought

Find the error pattern (what is correct, what is incorrect) in each of the following groups of solved problems. Then redo the problems correctly.

4 Name the fraction which represents the shaded part of each diagram:

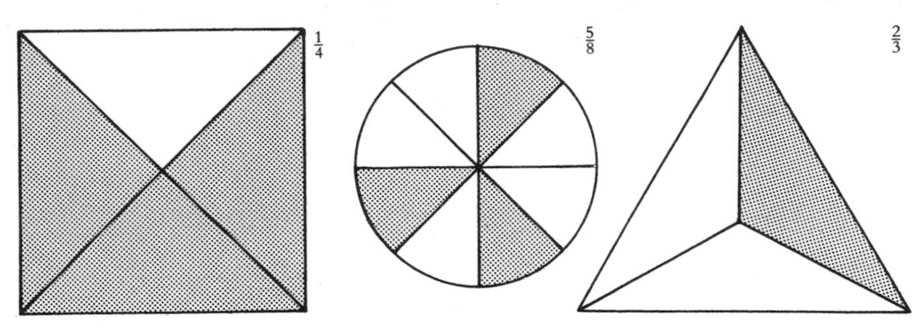

5 Name the fraction which represents the shaded part of each diagram:

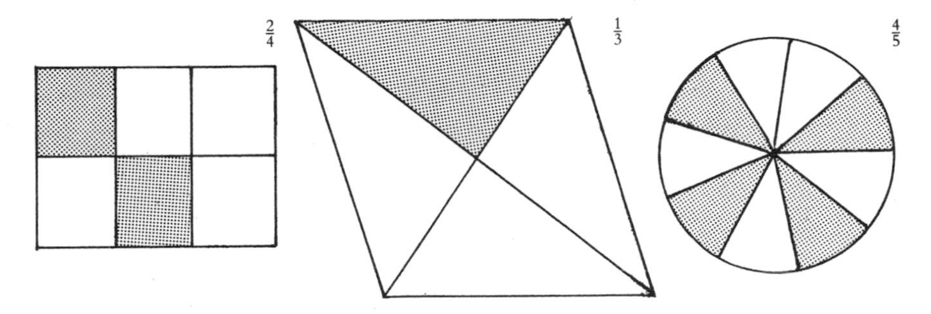

6 Create and solve a five-question quiz that reviews the major concepts learned in this chapter.

7 Skim the applications problems, choose the one that most interests you, and briefly write why.

8 What is the point of the following cartoon and how is it related to fractions (*New York Times*, 17 November 1985)?

Applications

9 Represent the length of each nail using fractions.

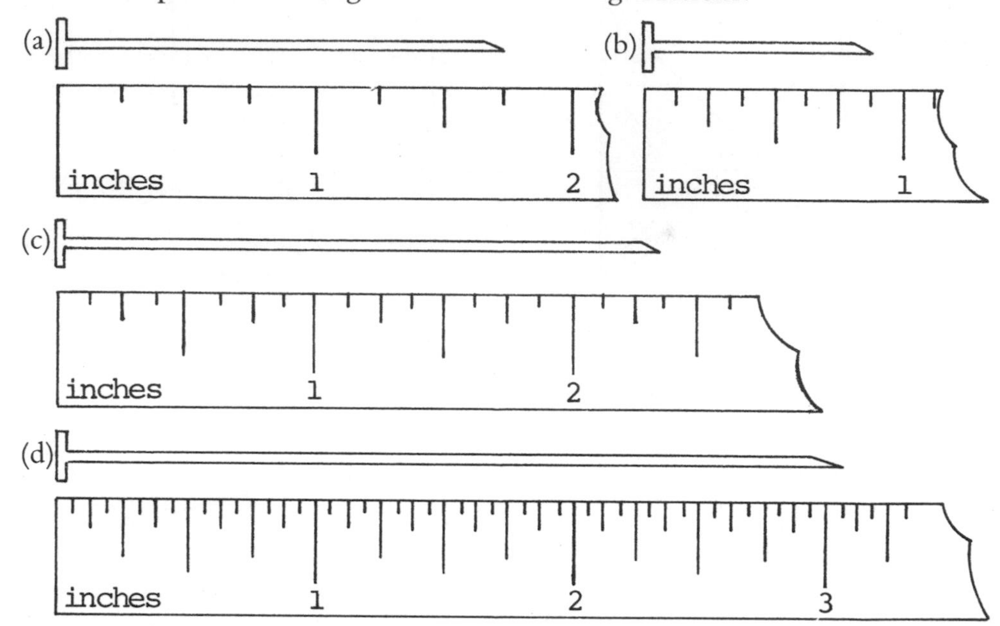

10 Read the following excerpt from 'UK's immigration laws: are they fair?' (*Labour Research*, November 1987). Summarize the main point and rewrite each of the main ratios as fractions.

'A report published in September 1987 by the British Refugee Council showed that the UK had 80 people seeking asylum per million inhabitants. Sweden by contrast had 1,598, Denmark 1,456, Switzerland 1,333, West Germany 1,117 and Austria 988 . . . In comparison with the UK, which has attempted to keep refugees out, India, a country with many fewer resources, has given refuge to 135,000 Sri Lankan Tamil refugees since 1983 – more than twice the number admitted by all the Western governments put together. As a result of neighbouring wars, many of the developing countries have been obliged to admit refugees. In Iran the ratio of refugees to population is 1:22, in Sudan it is 1:25, in Pakistan 1:36 and in Jordan 1:4.'

11 According to the following article (*In These Times*, 8 September 1982), what fraction of the nuclear power plant mishaps were particularly serious?

'No safety in numbers

'More than 4,000 mishaps occurred at US nuclear power plants in 1981 and over 83,000 atomic workers received radiation doses, according to a study conducted by Ralph Nader's Critical Mass Energy Project. Using more than 300 Nuclear Regulatory Commission (NRC) documents obtained through the Freedom of Information Act, the study found that of the 4,060 mishaps, 140 were especially serious. "These findings show that the day-to-day operation of nuclear reactors is plagued by equipment failures, human errors and design defects. The multitude of serious safety problems is a sure sign that without major changes it is just a matter of time before another serious accident occurs," the report said.

'But safety is in the eye of the beholder. The Atomic Industrial Forum, the nuclear industry's major trade association, responded that the industry has an "exemplary safety record".'

12 Use the passage below to answer these questions:

(a) What is Helen Keller's main point?
(b) How do the numbers she uses support her point?
(c) Why does she sometimes use fractions and at other times use whole numbers?

'Although Helen Keller was blind and deaf, she fought with her spirit and her pen. When she became an active socialist, a newspaper wrote that "her mistakes spring out of the . . . limits of her development". This newspaper had treated her as a hero before she was openly socialist. In 1911, Helen Keller wrote to a suffragist in England:

"You ask for votes for women. What good can votes do when ten-elevenths of the land of Great Britain belongs to 200,000 people and only one-eleventh of the land belongs to the other 40,000,000 people? Have your men with their millions of votes freed themselves from this injustice? (Zinn, 1980, p. 337)

13 Using the table on page 106 answer the following questions.

(a) Briefly describe the kinds of information contained in this chart.
(b) What general conclusions can you draw about the composition of the Senate in the ninety-fifth Congress?
(c) What fraction of all the senators were white?
(d) What fraction of all the senators were women?
(e) What fraction of the Democratic senators were not married?
(f) What fraction of the unmarried senators were Republican?
(g) Use the following updated information about the composition of the ninety-ninth Senate in 1985 to create and solve four fraction problems:

Male	98
Female	2
Black	–
Married	92

Senate Members of the ninety-fifth Congress, 1977

Characteristic	Senators		
	Total	Democrat	Republican
Total	★100	61	38
Male	★100	61	38
Female	–	–	–
White	★ 96	59	36
Black	1	–	1
Other	3	2	1
Married	★ 91	55	35
Not married	9	6	3

– represents zero. ★ includes 1 Independent.

Source: Compiled by the US Bureau of the Census from Congressional Quarterly Inc., Washington, DC, *Congressional Quarterly Almanac*; and US Congress, Joint Committee on Printing, *Congressional Directory*.

14 Write the fraction that represents the following situation:
(a) Each person's share when $18 is being equally shared among 4 people.
(b) Each person's share when 14 apples are being equally shared among 3 people.

15 Read below, 'A right to the tree of life', by Michael Robin (*The Nation*, 9 June 1984).
(a) What is the main point?
(b) How do the numbers support the main point?
(c) How are the numerical arguments related to the concept of fractions?

'In the 1920s demonstrators protesting the lack of decent health care for poor mothers and their infants marched on the Capitol in Washington with banners that proclaimed, "A Baby Saved Is a Citizen Gained". That cry badly needs to be resurrected today.

'While the status of fetuses holds center stage in American politics, the right to life has not yet been extended to thousands of black infants who die needlessly every year. Today, a black infant is twice as likely as a white infant to die before his or her first birthday. The most recent national infant mortality data broken down by race show that in 1981, 20 black babies per 1,000 died in their first year, as opposed to 10.5 white babies per 1,000. Had the black rate in 1981 been the same as the white, 5,584 of the 11,756 black infants who died that year would have lived.

'As the overall US infant mortality rate continues to decline – it was 11.2 per 1,000 in 1982 – the rate for blacks in some areas of the nation is rising, exceeding that of many Third World countries. In 1981 the infant mortality rate in black neighborhoods of Chicago was as high as 55 per 1,000; in central Harlem it was 28 per 1,000; and in parts of Baltimore, 59.5 per 1,000. In addition to numerous communities, thirteen states reported that the rate for blacks had increased in 1982. Most of those states are in the South, where infant death rates have been about 20 percent higher than in the rest of the country.

'Why are black babies so much more vulnerable than white babies? According to public health officials, the main reason is that the percentage of black newborns designated as low-birth-weight babies (under 5.5 pounds) is twice that of whites. As Myron Winick, professor of nutrition and pediatrics at Columbia University put it, "Pound for pound, the poor baby does as well as the rich baby; black babies do as well as white babies. The difference in mortality can be explained entirely by the fact that babies from these disadvantaged groups weigh on the average half a pound less at birth than middle-class babies."

'Many underweight babies are born to mothers who haven't had access to good prenatal care. Most public-health officials acknowledge that an expectant mother who receives no care is three times as likely to have a low-birth-weight baby as one who sees a doctor regularly. A recent study by the Children's Defense Fund reported that about 10 percent of all pregnant black women in the United States don't get prenatal care. In 1979, 386 Detroit women, or 1 percent of those who gave birth in the city that year, did not see a doctor until the day of delivery. Among their babies the death rate in the first year was a shocking 88 per 1,000. And in Washington, DC, the city with the highest infant mortality rate in the country – 20.3 per 1,000 in 1982 – 21 percent of nonwhite women receive inadequate medical attention while they are pregnant.

'Despite its pro-family rhetoric and its stated concern for the unborn and newly born, the Reagan Administration has severely cut many of the programs that specifically benefit pregnant women and infants. Federal budget cuts in Title V maternal and child health programs have meant that hundreds of poor women and their children have been turned away from prenatal and maternity services. About 90 percent of the Maternal and Infant Care clinics in ten states surveyed by the Children's Defense Fund had resources either cut or frozen in 1982. In Lexington County, Kentucky, the number of women receiving no prenatal care rose from 32 per 1,000 in 1980 to 55 per 1,000 in 1982, partly because the clinics had to refuse hundreds of needy women.

'In addition, the Administration has frozen funds for the Special Supplemental Food Program for Women, Infants and Children, better known as WIC. Although 9 million people are eligible for the program nationwide, the budget allows enough money for only about 2.5 million. According to a number of studies, WIC has been extraordinarily successful in reducing the incidence of low-birth-weight babies and has saved taxpayers' money. A Harvard University study shows that every $1 spent on nutritional supplements saves $3 in medical costs later for the care of a low-birth-weight baby.

'Federal programs to improve maternal and child health have also been hampered because they are administered by the states, where eligibility standards vary widely; as a result, many poor mothers and children have been excluded. In 1984, to be eligible to receive Aid to Families with Dependent Children (AFDC), a family of four in Oregon could have an annual income of up to $7,362; a similar family in Texas could earn no more than $3,618. Overall, the Southern states have the lowest eligibility ceilings and the lowest benefits. They also have the greatest number of poor mothers and children – and, not coincidentally, the highest infant mortality rates.

'Moreover, with Federal cuts reducing their budgets in the last few years, most states have tightened welfare eligibility requirements or have not raised income standards to keep pace with inflation, leaving thousands of poor women of childbearing age with no health insurance: in twenty states, women who do not qualify for AFDC automatically lose benefits under Medicaid as well. In 1982 the Reagan Administration further limited access to prenatal care by effecting regulations that disallow Federal AFDC reimbursements to state programs for women who are pregnant for the first time. Previously thirty states provided welfare benefits for first-time pregnancies; now only six provide assistance throughout a first pregnancy, and thirteen only after the sixth month.

'Any initiatives to expand health services to poor women and their children would almost certainly bring cries of "budget busting" from the Reagan Administration, but the simple truth is that universal access to services would be more equitable – and cost-effective. While it costs a total of about $1,500 to $2,000 to provide prenatal and delivery services to a pregnant woman, it costs more than $1,000 a day to provide intensive-care services for a premature or low-birth-weight baby. Fewer low-birth-weight babies would also mean fewer children who are retarded or are delayed in their development, which would also save millions of dollars in special care.

'It is time progressives took up the right-to-life banner, but for a different purpose. We should make sure that the Medicaid maternal and child health reform bill – a modest effort to extend coverage to poor women and children not eligible for Medicaid – gets adequate funding from the House-Senate conference committee that will soon be considering it. And we should press this year's Democratic Presidential candidate to put maternal and infant health high on the nation's political agenda. Unlike many issues that get far splashier press, this is a matter of life and death.'

Recreation

16(a) Theoretically, to cut a cake into two pieces, so that each of two people is satisfied that s/he has a fair share, have one person cut and the other choose. Why does this work?

(b) How can three people divide a cake into three pieces with which they are all satisfied?

Project

17 'Books for Bluefields', co-ordinated by Nan Elsasser and Patricia Irvine, is a project in which middle-school students in Albuquerque, New Mexico, wrote learning pamphlets for children who live on the (English-speaking) Atlantic coast of Nicaragua. The kids' writing was treated seriously, with letters from Elsasser and Irvine indicating, for example, that their books were 'accepted with the following revisions suggested'. The kids learned about Central America, as well as raising money to publish and distribute the books, which were actually delivered to the kids in Nicaragua.

Here is an excerpt, written by Cristián Cabrera, to introduce the concept of fractions. Create your own learning pamphlet for a friend who would like to study fractions.

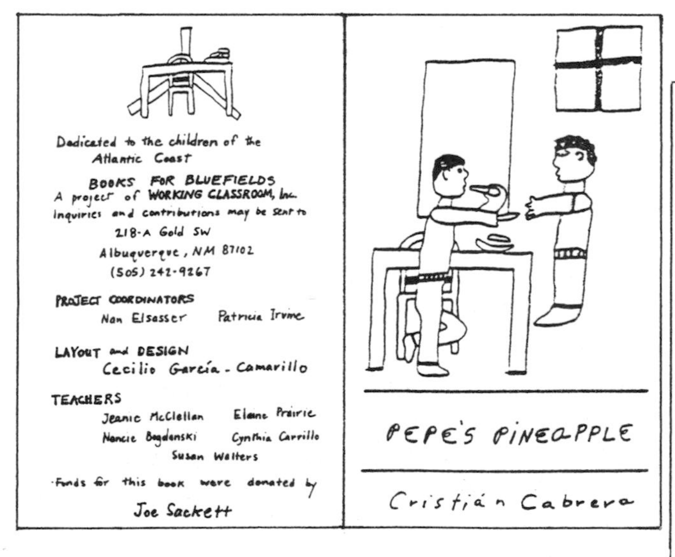

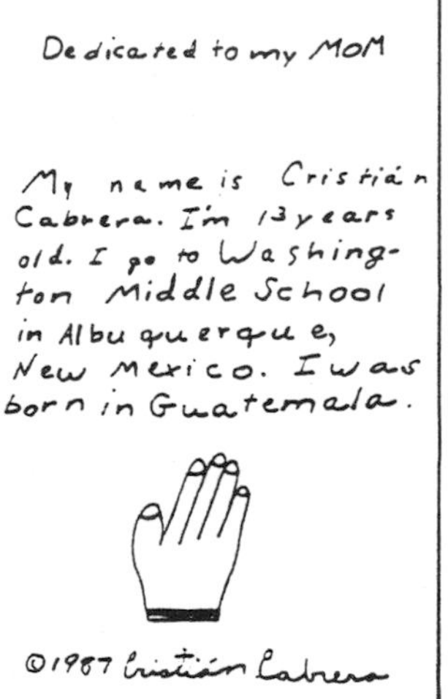

Pepe was hungry when he got home from School. He saw 3 cut pieces of round, juicy pineapple.
Knock Knock..
Pepe was about to get a piece when there was a knock on the door. Pepe opened the door and in came Paco.
Paco saw the pine-apple on the table and his mouth became watery. Pepe gave a piece to Paco and got one for himself.
Paco and Pepe finished at the Same time. They both eyed the last piece. They decided to cut it in 2 pieces.

"NO, like this," said Pepe. "The big piece for me and the small for you."
"NO! It's not good either," said Paco, "Like this. The big for me and the small for you."
Pepe said, "Let's cut it like this. The small piece for you and the big for me.
"NO that's not good," said Paco. "Cut it like this. The big piece for me and the small one for you.

These ways didn't work. One piece was always smaller than the other. They both wanted the big pieces.
"We must cut it in half," said Paco. Then both pieces were the same size. Each one would get an equal share. They were about to cut it when Juan came in and asked for a piece.
"You can't have any because there are only 2 pieces," said Pepe. Now Juan was smarter than Pepe and Paco. "Cut it in thirds and we can have 3 equal pieces. Now let's eat," said Juan. Juan cut the pineapple. They ate it happily, knowing that no one was cheated.

REVIEW QUIZ (the meaning of fractions)

(*Answers at end of chapter*)

1 Name the fraction represented by the shaded portion of the following diagram.

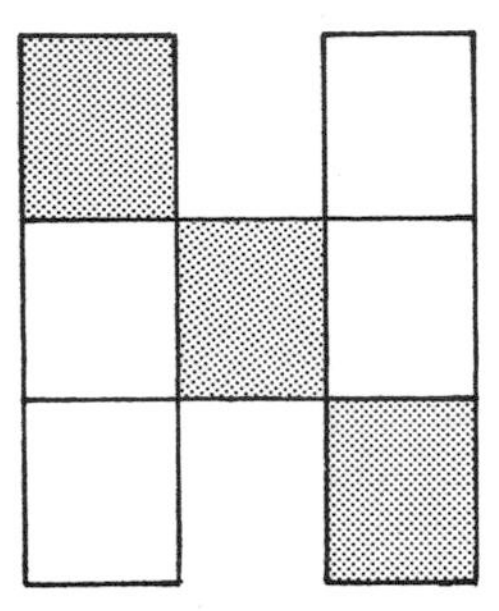

2 Shade in $\frac{2}{3}$ of the following diagram.

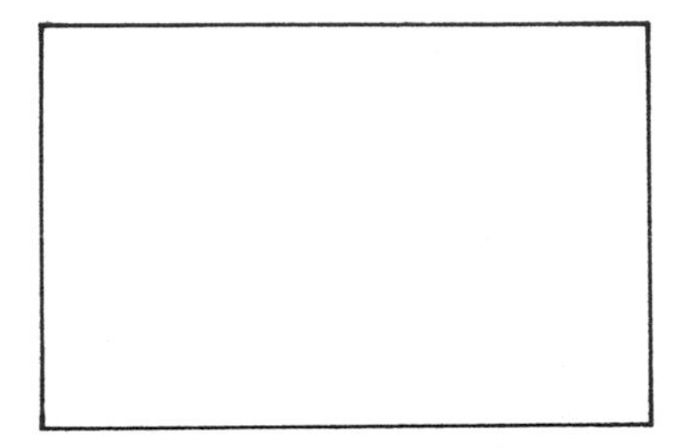

3 Represent the length of the following nail.

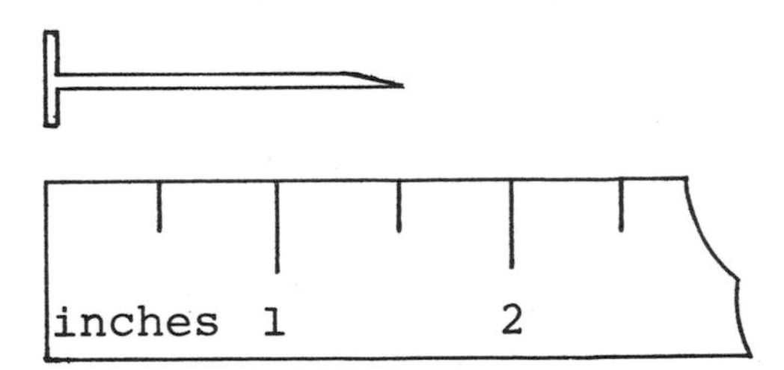

4 For these questions use the table from problem 13 earlier in this chapter.

(a) What fraction of the senators in the ninety-fifth Congress were married?

(b) What fraction of the Republican senators in the ninety-fifth Congress were non-white?

ANSWERS: CHAPTER 10

Solved problems ('you try these')

1 $\frac{4}{9}$

2

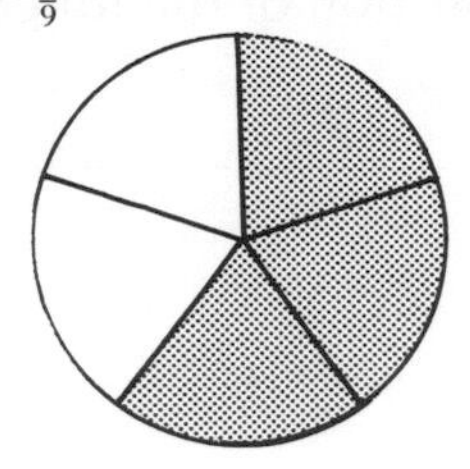

3 Here the person divided the whole into equal parts and shaded the correct number of them. But, s/he divided the whole into five, instead of four (the denominator) equal parts.

4 $\frac{7}{8}$ in

5 $$\frac{\text{number of black senators}}{\text{number of senators}} = \frac{1}{100}$$

6 $\frac{8}{7} = 1\frac{1}{7}$ pizzas

Unsolved problems

2 (a) $\frac{4}{8}$ (b) $\frac{4}{8}$ (c) $\frac{4}{12}$ (d) $\frac{4}{4} = 1$, since the entire figure is shaded.

3 (a)

(b) Or any five squares that you might have shaded.

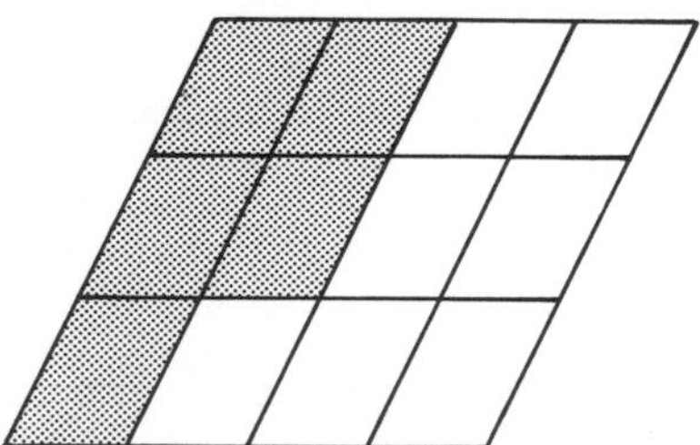

(c) Since this whole is divided into eight equal parts, you first make four equal groupings of two parts each, and then shade one of those groupings. This diagram illustrates that $\frac{2}{8} = \frac{1}{4}$.

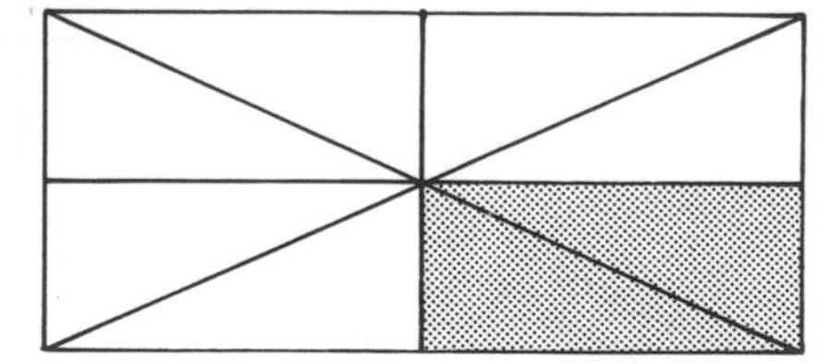

(d)

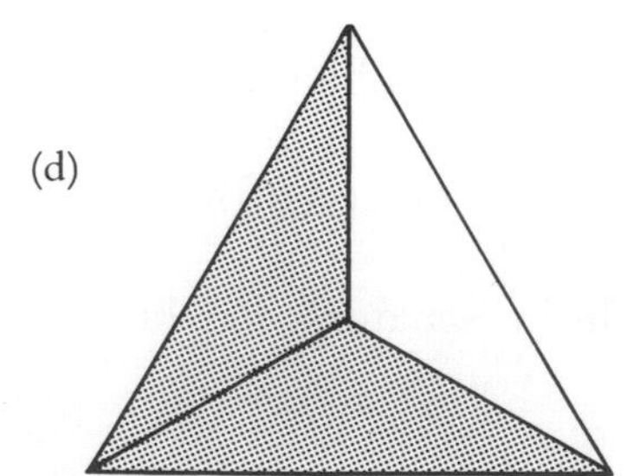

4 The solver wrote the fraction that represented the *un*shaded part of each diagram. The correct answers are: $\frac{3}{4}$, $\frac{3}{8}$, $\frac{1}{3}$.

5 The solver wrote the ratio of shaded to *un*shaded parts, the question asked for the ratio of shaded to *total* parts. The correct answers are: $\frac{2}{6}$, $\frac{1}{4}$, $\frac{4}{9}$.

8 The first frame implies each country will cut their nuclear arms in half, each keeping only one of the two equal halves. The second frame uses both countries as a total, each having $\frac{1}{2}$ the total, saying we'll keep our half, they can cut their entire amount.

9 (a) $1\frac{3}{4}$ in (b) $\frac{7}{8}$ in (c) $2\frac{3}{8}$ in (d) $3\frac{1}{16}$ in

10 Per million population, the UK has far fewer refugees seeking asylum than other European countries; moreover, the much poorer, developing countries have admitted larger numbers of refugees than all the Western governments combined. According to the article, 'The treatment of these and other refugees led the human rights organization, Amnesty International, to report in September 1987 that many governments are more concerned about denying sanctuary to political prisoners than they are about stopping persecution. The UK was cited as one of the countries which has become more restrictive.'

The fractions are:

$$\frac{80}{1{,}000{,}000} \quad \frac{1{,}598}{1{,}000{,}000} \quad \frac{1{,}456}{1{,}000{,}000} \quad \frac{1{,}333}{1{,}000{,}000} \quad \frac{1{,}117}{1{,}000{,}000} \quad \frac{988}{1{,}000{,}000}$$

$\frac{1}{22}$ $\frac{1}{25}$ $\frac{1}{36}$ $\frac{1}{4}$

11 $\frac{140}{4{,}060}$. In chapter 6, we'll see we can approximately reduce, or round, this fraction:

$$\frac{140}{4{,}060} \doteq \frac{150 \div 50}{4{,}000 \div 50} = \frac{3}{80} \doteq \frac{1}{25}$$

So about 1 out of every 25 nuclear power-plant 'mishaps' were serious.

12(a) Fighting for women's right to vote is not that important because men can vote and are still not freed from injustice.

(b) She uses the numbers of land distribution as an illustration of injustice: a small fraction of the people own most of the land.

(c) Her argument might be made even more powerful by showing both figures as fractions:

$$\frac{\text{large landowners}}{\text{total population}} = \frac{200{,}000}{40{,}200{,}000} = \frac{2}{402} = \frac{1}{201} \doteq \frac{1}{200}, \text{ and}$$

$$\frac{\text{small landowners}}{\text{total population}} = \frac{40{,}000{,}000}{40{,}200{,}000} = \frac{400}{402} = \frac{200}{201} \doteq 1.$$

So, she could have said $\frac{1}{200}$th of the population owns $\frac{10}{11}$ths of the land, while the rest, almost the total population ($\frac{200}{201}$), owns only $\frac{1}{11}$th of the land.

13(a) This type of table is called a matrix chart. It lists characteristics of the senators both horizontally and vertically. Along the horizontal, the senators are described in total and broken down into Democrats and Republicans. The vertical breakdown, by sex, race, and marital status, is done for each of the horizontal categories. So the chart describes the sex, race, and marital status of the total Senate, of the Democratic senators, and of the Republican senators.

(b) The Senate was all male, very largely white and married, and a sizeable majority of the senators were Democrats.

(c) $\frac{96}{100}$

(d) $\frac{0}{100}$ = 0. Here you can see this makes sense since there were no women senators in 1977.)

(e) $\frac{\text{unmarried Democrats}}{\text{total Democrats}} = \frac{6}{61}$

(f) $\dfrac{\text{unmarried Republicans}}{\text{unmarried senators}} = \frac{3}{9}$

(In Chapter 12 you'll learn that $\frac{3}{9} = \frac{1}{3}$, so we would say that one third of the unmarried senators were Republican.)

14(a) $\frac{18}{4}$ (In Chapter 12 you will learn that $\frac{18}{4} = 4\frac{2}{4} = 4\frac{1}{2}$, so each person gets $4.50.)

(b) $\frac{14}{3}$ (In Chapter 12 you will learn that $\frac{14}{3} = 4\frac{2}{3}$.)

15(a) 'Despite its pro-family rhetoric and its stated concern for the unborn and newly born, the Reagan Administration has severely cut many of the programs that specifically benefit pregnant women and infants.' This is a major cause of low birth weight among black babies, which is responsible for the fact that 'a black infant is twice as likely as a white infant to die before his or her first birthday'.

(b) Just one example for each point in (a): About 90 percent of the Maternal and Infant Care Clinics in ten states had resources either cut or frozen in 1982; the percentage of black newborns designated as low-birth-weight babies (under 5.5 lb) is twice that of whites; in 1981, 20 black babies per 1,000 died in their first year, as opposed to 10.5 white babies per 1,000.

(c) Infant mortality rates are really fractions because they are comparing two groups:

$$\frac{\text{babies who died}}{\text{1,000 babies born}}.$$

In this case for blacks, in 1981, infant mortality was $\frac{20}{1{,}000}$ and for whites, $\frac{10.5}{1{,}000}$.

16(a) Because if the person cutting cuts too big a piece, the other person chooses that piece and so gets more than half, so it's in the cutter's interest to cut in two equal halves.

(b) Call the people *A*, *B*, and *C*. *A* cuts what s/he thinks is $\frac{1}{3}$ of the cake. *B* and *C* can trim this piece if either feels it is too big. The last person to touch the slice keeps it as his/her share. The remainder of the cake, including any cut-off pieces, is now divided between the other two people as in (a) (Gardner, p. 161).

Review quiz

1. $\frac{3}{7}$ 3. $1\frac{1}{2}$ in 4(a) $\frac{91}{100}$ (b) $\frac{2}{38}$

11 The meaning of decimal fractions

Our place-value notation can now be extended to represent numbers smaller than 1, or fractions. You are already familiar with the decimal point in the representation of money (dollars and cents), but in other contexts the decimal point may not make any sense to you. (I can never resist a corny pun!)

In order to understand what quantities numbers written with decimal points represent, we must extend the place-value chart that we first saw in Chapter 9. (It is understood that the chart can be extended as far as needed in either direction.)

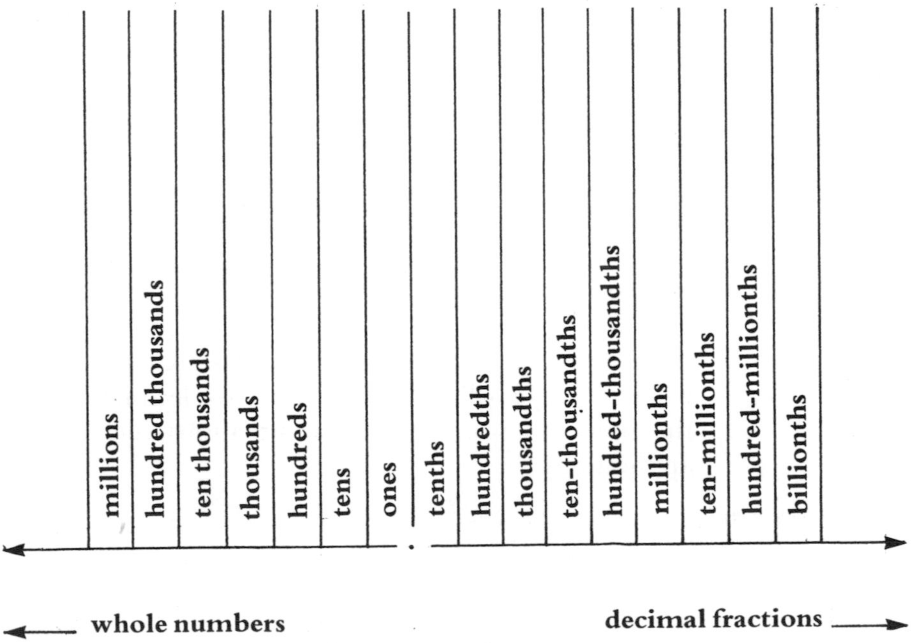

millions	hundred thousands	ten thousands	thousands	hundreds	tens	ones	.	tenths	hundredths	thousandths	ten-thousandths	hundred-thousandths	millionths	ten-millionths	hundred-millionths	billionths

← whole numbers　　　　decimal fractions →

SOLVED PROBLEMS

Drill

Problem 1 Write as a fraction, in words, each of the following numbers.
(Answers are below)

(a) 4.05 (b) 0.42 (c) 3.042

Solution In order to read and write mixed decimal fractions, we adopt the convention that the decimal point, separating the whole-number part from the fraction part, is read 'and'. Then the fraction part of the number is read as learned in Chapter 9, followed by the name of the last decimal place-value.

Answers
(a) 4.05 is written 'four and five hundredths'.
(b) 0.42 is written 'forty-two hundredths'.
(c) 3.042 is written 'three and forty-two thousandths'.

Notes

1 The 'ths' ending of the decimal-fraction place-values emphasizes that the places stand for fractions.
2 In order to make sure the decimal point is not ignored, decimal fractions are often written with a zero in the ones' place. 0.42 is just a more careful way of writing .42.
3 When there is no whole-number part, as in 0.42, the decimal point is not translated verbally as 'and'.
4 The zero between the decimal point and the four in 3.042 is also not translated directly. It is there to hold the places so that the 42 is in the correct position to represent 42 thousandths.
5 In the countries which reverse the usage of the comma and the decimal point (see Chapter 9, Solved Problem 1, n. 3), 3.042 is written '3,042' and instead of our '$3.25', in Haiti, say, where the currency is gourdes (G) and centimes (100 centimes = 1 G), 3 G and 25 centimes would be written '3,25G'. Remember that once you understand the concept of place-value and learn the actual decimal values of each place, it will become easy to switch to whichever notation is used in the country in which you live.
6 This text focuses on the decimal fractions of tenths, hundredths, and thousandths, which are the most common in everyday life. You come across decimal fractions in the millionths only in special situations such as science applications.

You try these *(Answers at end of chapter)*

(a) 6.02 (b) 0.57 (c) 12.057

Problem 2 Write each of the following decimals as fractions:

(Answers are below)

(a) 0.5 (c) 0.034

(b) 0.08 (d) 8.56

Solution In order to write decimal fractions as fractions, we use the number to the right of the decimal point as the numerator and the value of the last decimal place as the denominator.

Answers (a) $0.5 = \frac{5}{10}$ (c) $0.034 = \frac{34}{1{,}000}$

(b) $0.08 = \frac{8}{100}$ (d) $8.56 = 8\frac{56}{100}$

Note 8.56 represents 8 wholes and 56 hundredths, so it is written as a whole number with a fraction. We will learn more about this type of number, called a mixed number, in Chapter 14.

You try these *(Answers at end of chapter)*

(a) 0.9 (c) 0.067

(b)0.04 (d) 12.97

Problem 3 Write each of the following numbers in our place-value notation *(Answers are below)*

(a) fifteen and sixty-two thousandths

(b) seven and eight hundred and five thousandths

(c) fifty-six ten-thousandths.

Solution (a) 15.062.

Notes

1 The 'and' indicates the decimal point – the separation between the whole-number and fraction parts.

2 Make sure you use enough zeros so that the last digit of your number is in the correct decimal place. In this case we needed one zero to position the 2 in the thousandths' place.

(b) 7.805.

(c) 0.0056.

Note To check that you have the correct number of zeros, go through the place-values, starting with tenths, to make sure that the 6 is in the ten-thousandths place.

You try these *(Answers at end of chapter)*

(a) five and seventy-eight thousandths

(b) twelve and six hundred and twenty-one thousandths

(c) thirty-two ten-thousandths

Problem 4 Write in expanded notation *(Answers are below)*:

(a) 2.34

(b) 0.04
(c) 0.034

Solution In Chapter 9 we learned that the value of each place is equal to its digit multiplied by its place-value. For example,

in 2.34 the 2 represents 2 ones $= 2 \times 1 = 2$
in 3.24 the 2 represents 2 tenths $= 2 \times \frac{1}{10} = \frac{2}{10}$
in 3.42 the 2 represents 2 hundredths $= 2 \times \frac{1}{100} = \frac{2}{100}$

The value of the entire number is equal to the sum of the value of each place.

Answers (a) 2.34 = 2 ones + 3 tenths + 4 hundredths
$= 2 + \frac{3}{10} + \frac{4}{100}$

(b) 0.04 = 0 ones + 0 tenths + 4 hundredths
$= 0 + \frac{0}{10} + \frac{4}{100}$

(c) 0.034 = 0 ones + 0 tenths + 3 hundredths + 4 thousandths
$= 0 + \frac{0}{10} + \frac{3}{100} + \frac{4}{1{,}000}$

Notes 1 Another expanded notation would be to show the multiplication of the digit by its place-value. For example, $2.34 = (2 \times 1) + (3 \times \frac{1}{10}) + (4 \times \frac{1}{100})$.
2 2.34 can also be expanded as $2 + \frac{34}{100}$. (As we'll see in Volume 2, $\frac{3}{10} + \frac{4}{100} = \frac{34}{100}$.) However, this is not as detailed as the above expansions which treat each place-value separately.

You try these *(Answers at end of chapter)*
(a) 5.87 (b) 0.09 (c) 0.0048

Problem 5 Write each of the following as whole numbers
(Answers are below):
(a) 34 million.
(b) 3.4 million.
(c) 0.3 billion.

Solution Often, to save space, very large numbers are not written out with all their zeros. They are usually shortened to some number of millions or billions. If the number is not a whole number of millions or billions, a decimal point is used to indicate the fraction of millions or billions.

Answers (a) 34 million is written exactly as it reads: 34 million = 34,000,000.
(b) Technically, 3.4 million is 3 million and $\frac{4}{10}$ of a million.

Although you need to know how to multiply decimals (which you'll learn in Volume 2) fully to understand this problem, you can now become intuitively familiar with the fact that 3.4 million = 3,400,000. After all, it makes sense to write the 3 in the millions' place and it is reasonable to assume the 4 should be the next digit in the hundred-thousands' place.

(c) 0.3 billion is $\frac{3}{10}$ of a billion, or less than one billion. Following the pattern of part (b), we would write 0,300,000,000. Since this is a whole number, the zero in the billions place is not needed. So, 0.3 billion = 300,000,000. This seems 'tricky' because the answer is in the millions, rather than the billions (0.3 billion is a *fraction* of a billion, smaller than 1 billion).

You try these *(Answers at end of chapter)*

(a) 7 million (b) 6.7 million (c) 0.9 billion

Thought Problem 6 What is the relationship of each decimal fraction place-value to the place to its immediate right? *(Answer is below)*

Solution As with the whole-number place-values, each decimal fraction place is ten times larger than the place to its immediate right. For example,

$\frac{1}{10} = 10 \times \frac{1}{100}$ and $\frac{1}{100} = 10 \times \frac{1}{1,000}$.

This will not be totally clear to you until you study multiplying fractions in Volume 2 where you'll learn that $10 \times \frac{1}{100} = 10 \times 0.01 = 0.1 = \frac{1}{10}$. This is another instance (as in problem 5 above) where you have to temporarily tolerate some ambiguity. Right now, just begin to get the feeling that the place-values are related to each other by groupings of ten.

Answer Each decimal place-value (whether whole number or fraction) is ten times larger than the place-value to its immediate right.

You try this *(Answer at end of chapter)*

What is the relationship of each decimal fraction place-value to the place to its immediate left?

Applications Problem 7 Write the number in the following sentence in words, as a whole number: 'According to the Children's Defense Fund, as of 1976 there were 18.2 million children under 17 in the USA who had never visited a dentist.'

Solution As in Chapter 9, problem 5, I include these exercises to emphasize that too often we skip over numbers when we read, and therefore do not get the full impact of the statement. In Volume 2 we shall focus more clearly on the meaning of gigantic numbers. Here we're just getting practice reading them: 18.2 million = 18,200,000.

Answer As of 1976 there were eighteen million, two hundred thousand children under 17 who had never visited a dentist.

You try this *(Answer at end of chapter)*

As of March 1978, there were 0.9 million children under 18 with both parents unemployed.

Note One of my former students (Norma Corey) who is now a teacher pointed out to me that when numbers showing inequality or injustice are so large, there is a danger of people shrugging their shoulders and saying, 'How can I possibly fight to change that? If it were one child who lived on my block, maybe I could take that child to the dentist, but how can I take 18,200,000 children to the dentist?' I answered that, first, it is important to understand the numerical extent of the injustice in our society. If only a few children had not been to a dentist, we could conclude it was because their parents were negligent or they were scared of the dentist, that is, we could conclude it was a personal problem. When so many millions

of children are involved, it is clearer that it is not a personal problem, but that there is something wrong with the way our society is organized. Second, one individual alone cannot change the way society is organized, but individuals can work together to change the world. At the end of this book is a list (see Appendix) of the addresses of organizations you might wish to contact to work on the issues mentioned in this text; it also contains suggestions for how to form your own community organization to work on problems specific to your community, and how your knowledge of maths will help in your struggle.

UNSOLVED PROBLEMS

(*Answers at end of chapter*)

Drill

1 Write each of the following numbers in words:

(a) 3.48 (b) 0.348 (c) 9.006 (d) 9.0006

2 Write each of the following decimals as fractions:

(a) 0.09 (b) 0.347 (c) 15.06 (d) 12.012

3 Write each of the following numbers in our place-value notation:
(a) seven and fifty-eight hundredths
(b) seven and fifty-eight thousandths
(c) four hundred and nine thousandths
(d) six hundred and seven ten-thousandths

4 Write in expanded notation:
(a) 6.05 (b) 0.042

5 Write each of the following as whole numbers:
(a) 49 billion (b) 29.3 million (c) 2.93 million

Thought

6(a) Describe a situation in which you would want to use fractions instead of decimals.
(b) Describe a situation in which you would want to use decimals instead of fractions.
(c) Discuss the advantages and disadvantages of representing numbers as fractions or as decimals.

7 What's funny about the following cartoon and how does the humour relate to the maths you've studied so far in this text?

8 Find the error pattern (what is correct, what is incorrect) in the following group of solved problems. Then redo the problems correctly.

Write each of the following numbers in our place-value notation:

thirty-nine hundredths = 0.039
five tenths = 0.05
seventeen thousandths = 0.0017

9(a) Create and solve a quiz that reviews the major concepts learned in this chapter.

(b) Which of these questions interests you the most? Write about why.

Applications

Write the numbers in each of the following sentences in words, as whole numbers:

10 A June 1981 letter from the War Resisters League stated that, over the next five years, the Pentagon expects to spend $1.3 trillion on instruments of destruction. The letter goes on to indicate that for the cost of just two fewer F-14 fighter planes, all of Reagan's 1981 cuts in child nutrition programmes could be completely restored.

11 A 1 January 1981 issue of *WIN* magazine lists the 100 companies which got the largest military contracts from the government in 1979:

(a) General Dynamics heads the list with $3,492.1 million.

(b) General Electric is fourth with $2,042.5 million, and this military business represents only about $\frac{1}{10}$ of all GE's sales!

(c) Other companies whose major business is elsewhere still made the top 100 list. American Telephone and Telegraph (AT&T), for example, is 18th with $569.6 million. This represents only about $\frac{1}{100}$ of AT&T's sales.

(d) The smallest contract awarded in the top 100 was to Coastal Drydock Repair for $76.2 million.

12 Study the table below.

Employment in Great Britain (millions)

		Manufacturing	Services	Other*	Total
1979	June	7.13	13.21	2.29	22.64
1986	March	5.21	14.04	1.83	21.07
	June	5.15	14.13	1.82	21.08
	September	5.10	14.21	1.80	21.10
	December	5.09	14.27	1.79	21.15
1987	March	5.05	14.37	1.80	21.21
	June	5.06	14.47	1.80	21.30
	September	5.03	14.54	1.79	21.36
	December	5.03	14.67	1.77	21.47

*Mainly energy, water, construction, agriculture.

(a) What is the main point of the table?

(b) Write the total employment numbers as whole numbers.

13 Look through a newspaper or magazine for various uses of decimal fractions and write each of the uses you find in words.

REVIEW QUIZ (the meaning of decimal fractions)

1 Write in words as a decimal fraction: 9.08.

2 Write in decimal place-value notation: ten million and fifty-six thousandths.

3 Write as a fraction: 0.045.

4 Write the numbers in each of the following sentences in words, as whole numbers:

(a) According to the Children's Defense Fund, in 1976 there were 1.3 million children under 17 who had never visited a doctor.

(b) In 1979, IBM had a military contract for $552.6 million, which represented about $\frac{2}{100}$ of IBM's total sales.

ANSWERS: CHAPTER 11

Solved problems ('you try these')

1 (a) six and two hundredths
(b) fifty-seven hundredths
(c) twelve and fifty-seven thousandths

2 (a) $\frac{9}{10}$ (b) $\frac{4}{100}$ (c) $\frac{67}{1,000}$ (d) $12\frac{97}{100}$

3 (a) 5.078 (b) 12.621 (c) 0.0032

4 (a) $5 + \frac{8}{10} + \frac{7}{100}$
(b) $\frac{0}{10} + \frac{9}{100}$
(c) $\frac{0}{10} + \frac{0}{100} + \frac{4}{1,000} + \frac{8}{10,000}$

5 (a) 7,000,000 (b) 6,700,000 (c) 900,000,000

6 Each decimal place-value (whether whole number or fraction) is ten times smaller than the place-value to its immediate left.

7 0.9 million = 900,000 = nine hundred thousand

Unsolved problems

1 (a) three and forty-eight hundredths
(b) three hundred and forty-eight thousandths
(c) nine and six thousandths
(d) nine and six ten-thousandths

2 (a) $\frac{9}{100}$ (b) $\frac{347}{1,000}$ (c) $15\frac{6}{100}$ (d) $12\frac{12}{1,000}$

3 (a) 7.58 (b) 7.058 (c) 0.409 (d) 0.0607

4 (a) $6 + \frac{0}{10} + \frac{5}{100}$
(b) $0 + \frac{0}{10} + \frac{4}{100} + \frac{2}{1,000}$

5 (a) 49,000,000,000
(b) 29,300,000
(c) 2,930,000

6 (a) Measuring with a ruler calibrated in inches (can you think of others?)
(b) Calculating with dollars and cents (although to save space, money amounts are written as fractions on the stock-market page of a newspaper, where for example '+$3\frac{4}{4}$' means the selling price of the stock went up $3.75 from yesterday's selling price).

7 The cartoon is an ironic commentary on environmental pollution – the town is proud to advertise that it *only* has 1.003 parts per trillion of these horrible pollutants! The relationship to decimals is the number 1.003 = $1\frac{3}{1,000}$; the relationship to fractions is the *rate* of pollution.

$$1.003 \text{ per trillion} = \frac{1.003}{1,000,000,000,000}$$

8 The solver correctly wrote the number and clearly knows something about the decimal fraction place-values because s/he was only wrong by one extra place in each problem. Some people get confused by the fact that the ones' whole-number place has no matching decimal-fraction place, even though all the other places do (for example, tens and tenths, thousands and thousandths). This person probably thought the first decimal fraction place was oneths because if that were so, his/her answers would be correct. For example, *if* the decimal fraction places were oneths, tenths, hundredths, and so on, then the 9 in 0.039 would be in the hundredths place. But the first decimal fraction place is tenths, so the correct answers are: 0.39; 0.5; 0.017.

10 The Pentagon is expected to spend one trillion, three hundred billion dollars on instruments of destruction over a five-year period beginning in 1981.

11 (a) General Dynamics' military contract was for three billion, four hundred and ninety-two million, one hundred thousand dollars. (This number is written in the millions, rather than as 3.4921 billion, because in the table all the contracts are listed in millions and most are actually in the millions, not the billions.)
(b) GE's military contract was for two billion, forty-two million, five hundred thousand dollars, representing only about one-tenth of their sales.
(c) AT&T's military contract was for five hundred and sixty-nine million, six hundred thousand dollars, representing only about one-hundredth of their sales.
(d) Coastal Drydock Repair's military contract was for seventy-six million, two hundred thousand dollars. (If the chart had been listed in billions rather than millions, this contract would have been listed as 0.0762 billion.)

12 (a) The table shows that during 1979–87 the number of manufacturing jobs has been declining and the number of service jobs increasing; it also shows that the total number of jobs increased from 1986–7, but not to the 1979 level. According to the notes accompanying the table, more than half the increase in 1987 was in part-time jobs.
(b) 22,640,000; 21,070,000; 21,080,000; 21,100,000; 21,150,000; 21,210,000; 21,300,000; 21,360,000; 21,470,000.

Review quiz

1 nine and eight hundredths.

2 10,000,000.056.

3 $\frac{45}{1,000}$.

4 (a) In 1976, one million, three hundred thousand children under seventeen had never visited a doctor.
(b) IBM's military contract was for five hundred and fifty-two million, six hundred thousand dollars, representing only about two hundredths of their sales.

12 Comparing the size of whole numbers and decimal fractions

The following table from SANE's *Nuclear Wastes Fact Sheet* lists the temporary storage areas for wastes from commercial uses of radioactive material, mainly from nuclear power plants.

Location	Amount of waste stored (in cubic feet)	
Barnwell, South Carolina	3,520,000	(low level)
Beatty, Nevada	1,970,000	(low level)
Hanford, Washington	510,000	(low level)
Maxey Flats, Kentucky	4,950,000	(low level)
Morris, Illinois	4,061	(spent fuel)
Sheffield, Illinois	2,400,000	(low level)
West Valley, New York	32,720	(high level)

Spent fuel is stored at the site of the 72 operating commercial reactors. Also, about 3 billion cubic feet of uranium tailings from commercial and military uses are stored at over 130 mill sites, predominantly in Colorado, New Mexico, Utah, and Wyoming, and tailings have even been left out in large open piles, some as close as a few hundred feet to urban thoroughfares.

Source: Department of Energy, *Report to the President by the Interagency Review Group on Nuclear Waste Management*, March 1979.

These wastes include long-lived and extremely toxic elements, such as plutonium. They emit ionizing radiation which can impair or destroy living cells and lead to cancer or genetic damage. These sites are temporary because these wastes are dangerous to humans for hundreds of thousands of years, and no permanent storage solution has yet been found.

I hope that, as you become more comfortable using numbers, a chart like this will provoke many questions, such as 'Why are we continuing to produce more waste?', 'Specifically, how large is the amount of waste currently stored?' and 'Where is the most waste stored?' The first question can only be answered after studying the history of the nuclear-power industry and the details of how profits have been considered before people; the mathematics needed to answer the second question will be explored in Volume 2; the mathematics of the third question is the topic of this chapter.

Whenever we have a list of two or more numbers, we can ask and answer questions about which number is largest, and which is smallest, and we can put the whole list in size order. In many cases, this way of ordering numerical facts is helpful to our analysis of the situations represented by the numbers. Of course, the numbers alone don't present the complete picture. For example, although West Valley, NY, stores one of the smaller amounts of cubic feet of waste, its waste is 'high level' and therefore more dangerous than larger amounts of 'low level' waste.

You can probably tell intuitively that 4,950,000 is the largest number on the chart. I state a rule for comparing numbers because we have less intuition about comparing decimals such as 0.03 and 0.034, and because understanding rules for very basic maths forms a foundation for understanding how numbers work and for being able to look at numbers and know what to do with them in various circumstances. In general, to compare the size of two whole numbers or decimal fractions:

1 Compare the corresponding digits in each place-value, starting from the largest place-value.
2 If the numbers are not identical, they will eventually have different digits in a particular place-value.
3 The larger number is the one with the larger digit in that particular place-value.

Looking at the table, and using the rule, we start comparing in the millions' place and immediately see that Maxey Flats, Kentucky, has the largest digit in that place, so as of 1979 Maxey Flats had stored the largest amount of cubic feet of commercial radioactive waste.

SOLVED PROBLEMS

Drill

In problems 1–5 replace the question mark with the correct comparison symbol. *(Answers are below)*

Problem 1 (a) 34,234 ? 34,324
(b) 67,894 ? 7,894
(c) 1,235,469 ? 1,235,469

Solution There are three symbols used to compare numbers:

'=' means 'is equal to' (and '≠' means 'is not equal to')
'<' means 'is less than'
'>' means 'is greater than'

I remember the difference between < and > by the fact that the smaller opening of the arrow always points to the smaller number:

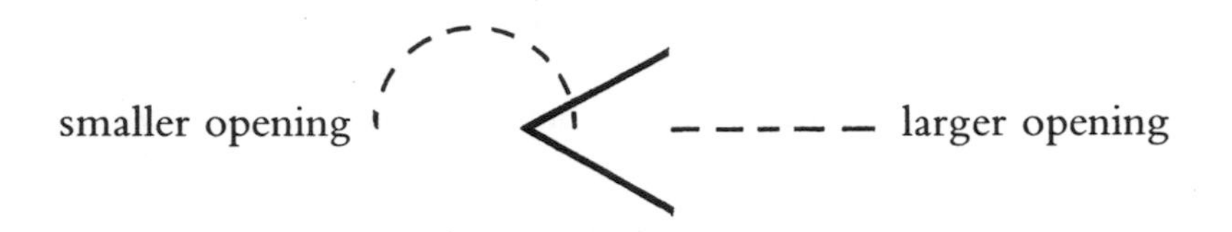

5 is less than 10   10 is greater than 5
5 < 10   10 > 5

In both cases, the smaller part of the arrow points to the smaller number, 5.

Answers
(a) 34,234 < 34,324 because the first digit that differs is in the hundreds' place and the hundreds' digit of 34,234 is smaller than the hundreds' digit of 34,324.
(b) 67,894 > 7,894 because 7,894 has no ten thousands' digit, so any number with a digit in the ten thousands' place is larger than 7,894.
(c) 1,235,469 = 1,235,469 because the digits in each place-value are identical.

Note We can read these number comparisons from left to right, or from right to left. For example, we can say 34,234 is less than 34,324; or 34,324 is greater than 34,234.

You try these *(Answers at end of chapter)*
(a) 65,438 ? 65,483 (b) 125,126 ? 126,125 (c) 12,345 ? 2,345

Problem 2 4.256 ? 4.265

Solution 4.256 < 4.265 because the hundredths' digit of 4.256 (5) is smaller than the hundredths' digit of 4.265 (6).

You try this *(Answer at end of chapter)*

6.784 ? 6.847

Problem 3 6.4 ? 6

Solution Intuitively, since 6.4 represents 6 and something else, you can see that $6.4 > 6$. In order to solve this problem by following the rule, we must have a digit in the tenths' place of 6 to compare with the 4 in the tenths' place of 6.4. Any whole number is understood to have a decimal point immediately to the right of the ones' place and zeros in all the decimal places to the right of the decimal point. Therefore, $6 = 6.0$. (You already know this when it comes to money: $6 = $6.00, £6 = £6.00) $6.4 > 6.0$ because the tenths' digit of 6.4 is greater than the tenths' digit of 6.0.

Answer $6.4 > 6$

You try this *(Answer at end of chapter)*

7 ? 7.2

Problem 4 .805 ? 8

Solution Intuitively, any number whose only non-zero digits are to the right of the decimal point is smaller than 1. In this case, .805 represents $\frac{805}{1,000}$ a fraction smaller than 1, since $\frac{1,000}{1,000} = 1$. Because $0.805 < 1$, it follows that $.805 < 8$. In order to solve this problem by following the rule, use the fact that $.805 = 0.805$ and $8 = 8.000$. $0.805 < 8.000$ because the ones' digit of 0.805 is smaller than the ones' digit of 8.000.

Answer $.805 < 8$

Note This may seem counter-intuitive if you don't focus on the decimal point, since $805 > 8$. The location of the decimal point is crucial, since the decimal point tells us that .805 is $\frac{805}{1,000}$ Read carefully so you don't overlook the decimal point.

You try this *(Answer at end of chapter)*

9 ? .95

Problem 5 0.3 ? 0.30

Solution These two numbers are equal. For now you may have to settle for an intuitive understanding of fact, until we study reducing fractions in chapter 6.

$0.3 = \frac{3}{10}$

$0.30 = \frac{30}{100}$ and you may remember that $\frac{30}{100} = \frac{3}{10}$

If you are not familiar with reducing fractions, it may help to think of $\frac{3}{10}$ of a dollar (3 dimes) and $\frac{30}{100}$ of a dollar (30 pennies). Clearly, 3 dimes = 30 pennies.

In general, adding zeros in the places to the right of a number written with a decimal point does not change the value of the number. Intuitively this makes sense because the added zeros do not push the non-zero digits into other places. For example, 0.428 = 0.4280 = 0.42800. In all three cases, we have 4 tenths, 2 hundredths, and 8 thousandths. However, 428 does not equal 4,280 because the added zero changes 4 hundreds to 4 thousands, 2 tens to 2 hundreds, and 8 ones to 8 tens. Finally, just as the zeros can be added in the places to the right of a number written with a decimal point, they can be omitted. For example, 6.7800 = 6.78 and 0.0450 = 0.045.

Answer 0.3 = 0.30

You try this *(Answer at end of chapter)*

.52 ? .520

Problem 6 Put the following group of numbers in size order from largest to smallest:

374; 24,047; 24.36; 24,360; 30.04; 30.047; 0.5; 0.52.

Solution First, separate the numbers into groups with the same largest place-value:

24,360; 24,047	(ten thousands)
374	(hundreds)
24.36; 30.04; 30.047	(tens)
0.5; 0.52	(tenths)

Clearly, numbers in the ten thousands are larger than numbers in the hundreds, which, in turn, are larger than numbers in the tens, etc.

Next, compare the numbers in each group:

$$24{,}360 > 24{,}047$$

$$30.047 > 30.04 > 24.36$$

(Remember: 30.047 contains 7 thousandths and 30.04 contains 0 thousandths.)

$$0.52 > 0.5$$

(If it is clearer to you, add zeros when comparing decimals. 0.5 = 0.50, and comparing 0.52 and 0.50 clearly leads to 0.52 > 0.50.)

Answer 24,360 > 24,047 > 374 > 30.047 > 30.04 > 24.36 > 0.52 > 0.5

Notes 1 A shorthand way for writing $7 > 5$ and $5 > 3$ is $7 > 5 > 3$, which reads '7 is greater than 5, and 5 is greater than 3'. This shorthand can be extended to entire strings of numbers as in this answer.

2 We'll study comparing fractions in chapter 15. Basically, we first change fractions to decimals and then compare the decimal fractions.

You try this *(Answer at end of chapter)*

352; 0.352; 0.52; 35,235; 325; 35.2; 35,325; 35.25

Applications

Problem 7 Use the table at the beginning of this chapter to decide which site has the third largest amount of commercial waste stored.

Location	Amount of waste stored (in cubic feet)	
Barnwell, South Carolina	3,520,000	(low level)
Beatty, Nevada	1,970,000	(low level)
Hanford, Washington	510,000	(low level)
Maxey Flats, Kentucky	4,950,000	(low level)
Morris, Illinois	4,061	(spent fuel)
Sheffield, Illinois	2,400,000	(low level)
West Valley, New York	32,720	(high level)

Source: Department of Energy, *Report to the President by the Interagency Review Group on Nuclear Waste Management*, March 1979.

Solution There are four sites with at least one million cubic feet of waste stored. In size order,

$4{,}950{,}000 > 3{,}520{,}000 > 2{,}400{,}000 > 1{,}970{,}000$, so . . .

Answer The third largest amount, 2,400,000 cu. ft, is stored at Sheffield, Illinois.

Note The answer to this question is not totally clear, because the note at the bottom of the table does not indicate how much spent fuel is stored at the commercial reactors or how much uranium tailing is stored at each mill site. Some of those sites might have more than 4,950,000 cubic feet of waste stored. When you use numbers in real life, you often have to investigate further to clarify the information. Because this is a textbook problem, we'll just leave this answer with the note indicating what else must be determined in order to solve the problem more completely.

You try this *(Answer at end of chapter)*

Which site has the smallest amount of commercial waste stored?

UNSOLVED PROBLEMS

(Answers at end of chapter)

Drill

1 Replace the question mark with the correct comparison symbol:

(a) 3,352 ? 3,252
(b) 46,784 ? 49,784
(c) 2,475,235 ? 24,752,350
(d) 8.65 ? 8
(e) 9 ? .98
(f) 1 ? 1.002
(g) .349 ? 1
(h) .304 ? .3040

2 Put the following groups of numbers in size order from smallest to largest:

(a) 48; 804; 84; 8,004
(b) 12,432; 12,342; 1,342; 13,423

3 Put the following groups of numbers in size order from largest to smallest.

(a) 15; 13.2; 15.53; 13.04
(b) 0.003; 0.0003; 0.030; 0.30; 0.00003; 0.303

Thought

4 What's funny about each of the following cartoons and how does the humour relate to the main concepts studied in this text?

THE HISTORY OF THE ARMS REDUCTION TALKS
TO BE CONTINUED...
A FREEZE ON NUCLEAR WEAPONS WILL NOT ACHIEVE THE REAL ARMS REDUCTIONS THAT I SEEK
A FREEZE WOULD FIX OUR STRATEGIC FORCES AT ABOUT THIS LEVEL
HOWEVER, AN ARMS RACE WOULD BRING THEM UP TO THIS LEVEL-
MORE ROOM FOR REDUCTIONS!
WASSERMAN

5 Write briefly about the meaning you see in this story from Central Asian oral tradition (recounted in Ornstein, p. 17).

'Is There Any Number Higher Than 100?

'A man, having looted a city, was trying to sell an exquisite rug, one of the spoils. "Who will give me 100 pieces of gold for this rug?" he cried throughout the town.

'After the sale was completed, a comrade approached the seller and asked, "Why did you not ask more for that priceless rug?"

'"Is there any number higher than 100?" asked the seller.'

6 Create and solve a quiz that reviews the major concepts learned in this chapter.

7 Look back over the journal entries you have written about your understanding of the meaning of whole numbers, fractions, and decimal fractions. Write a list of what is currently clear to you and what questions you still have about the meanings of numbers.

Applications

8 An article in the June 1981 issue of *The Progressive* magazine lists the top twenty universities in military research contracts for fiscal year 1980. Put the following alphabetically ordered list in size place from largest to smallest military contract:

	$
California Institute of Technology	5,428,000
Carnegie Mellon University	7,335,000
Columbia University	4,848,000
Georgia Technical Research Institute	14,758,000
Harvard University	4,902,000
Illinois Institute of Technology	26,319,000
Johns Hopkins University	163,327,000
Massachusetts Institute of Technology	154,564,000
Pennsylvania State University	12,226,000
Stanford University	18,068,000
University of Alaska	8,119,000
University of California (system)	29,679,000
University of Dayton	13,859,000
University of Illinois	6,797,000
University of New Mexico	5,472,000
University of Pennsylvania	4,900,000
University of Rochester	15,480,000
University of Southern California	10,260,000
University of Texas	15,772,000
University of Washington	10,069,000

(The Appendix gives concrete steps you can take to determine how much military research is going on at your local university.)

9 Put the following activities in size order from the smallest to largest number of calories used per minute per pound.

Activity	*Calories per minute per pound*
Basketball:	
Moderate	0.047
Vigorous	0.066
Bicycling:	
Slow (5 miles per hour)	0.025
Moderate (10 mph)	0.05
Fast (13 mph)	0.072
Dancing:	
Slow	0.029
Moderate	0.045
Fast	0.064
Gardening	0.024
Running	
6 mph	0.079
10 mph	0.1
12 mph	0.13
Swimming (crawl):	
20 yards per minute	0.032
45 yds/min	0.058
50 yds/min	0.071
Tennis:	
Moderate	0.046
Vigorous	0.06
Walking:	
2 mph	0.022
3 mph	0.03
4 mph	0.039
5 mph	0.064

Source: Frank Vitale, *Individualized Fitness Programs*, Prentice-Hall, 1973.

10 Discuss any conclusions and any questions you have after comparing the numbers in the following table (Leiner, 1981, p. 204).

Enrolment, Schools, and Teachers in Cuba, 1958–9 and 1979–80

Type of	1958–9			1979–80		
education	students	schools	teachers	students	schools	teachers
Kindergarten	91,700			122,600		
Elementary	625,700	7,567	17,355	1,550,300	12,678	91,317
Secondary	88,100	80	4,571	1,150,300	1,606	80,188
Special				25,000	253	4,936
Adult				392,000		24,201
Higher	15,600	6		200,200	39	10,736
Others	3,700	28	669	12,500	142	1,300
Total	824,800		22,595	3,452,900		212,789
Students per 100 inhabitants	12			35		

Source: Cuban Ministry of Education.

Note: Kindergarten takes place in elementary school and day-care centres. Blanks indicate data wasn't available. 'Others' includes special programmes for thirteen- to sixteen-year-olds who are behind normal grade level and programmes for foreigners taking special courses.

11 Discuss the comparisons it makes sense to consider to understand the following statement.

'The relationship of these global corporations with the poorer countries was an exploiting one, it was clear from US Department of Commerce figures. Whereas US corporations in Europe between 1950 and 1965 invested \$8.1 billion and made \$5.5 billion in profits, in Latin America they invested \$3.8 billion and made \$11.3 billion in profits, and in Africa they invested \$5.2 billion and made \$14.3 billion in profits.' (Zinn, p. 557)

12 Read the following article (from the *Observer*, 29 June 1986). Create and solve two maths problems whose solution involves comparing numbers.

'Money for jam – but none for lentils.

PHILIPPA BRAIDWOOD
reports on why the poor don't eat healthily

'It's widely preached that it costs no more to eat healthily. This may be true for typical middle-class people, who can save money by eating cheap vegetables in place of expensive meat. But for the poor it's a different matter.

'A report published tomorrow by the London Food Commission claims a "NACNE diet" – one which conforms to the National

Advisory Committee on Nutrition Education guidelines – costs 35 per cent more than the amount currently spent on food by low-income families. It also says the part of the Supplementary Benefit allowance supposed to be for food falls short of the cost of the NACNE diet by more than 13 per cent.

'In a research project, 500 dietitians, who were neither extravagant nor penny-pinching, made a conscious effort to follow a NACNE diet. Their records revealed that in Britain today it costs £12.75 a week to eat healthily. The amount that DHSS staff are told to allocate for food when calculating an adult's benefits, however, is only £11.30. And for an eleven- to sixteen-year-old boy, who, says the DHSS, needs more to eat than an adult, the allowance is a mere £4.40 a week.

'In order to be healthy, we're told by NACNE and more recently by COMA (Committee on the Medical Aspects of Food Policy), we should eat more fresh fruits, vegetables, pulses and wholegrain cereals and fewer fatty and sugary foods. To make such changes without losing weight, we have to eat more: we might have to eat six oranges, at a cost of 60p, for instance, to get the same calories as from a 12p cake.

'Not surprisingly, those on low incomes already shop more efficiently than the better-off: for what they spend, they get more energy and nutrients, penny for penny – with the exception of vitamin C – than those with high incomes and big food bills.

'Yet the poor spend so much less on food than the better-off that they end up eating fewer nutrients. They spend only half as much on fresh fruit but more on fatty cuts of meat and calorie-dense foods such as pies, sausages, biscuits and jam. They eat low-fibre cereal – they would have to spend half as much again to get the equivalent energy from a high-fibre one. They buy white bread (wholemeal costs 25 per cent more) and full milk. The poor spend less on alcohol than the better-off, but their children are more likely to drink orange squash than nutritious fresh orange juice.

'At times when money's scarce, says the report, people with low incomes cut back on food expenditure, skipping meals altogether.

'Is it really sensible to advise them to buy more of foods they can't afford? The report says not.

'Dietitian Isobel Cole-Hamilton, co-author of the report, says: "I've been guilty of telling people it's possible to eat well cheaply for years. It's nonsense. If you're prepared to eat no meat, hardly any cheese and live mainly on pulses and cereals it's probably *possible* to get the NACNE composition right, in terms of amounts of fibre, fat and so on, but not to follow the guidelines on more fresh foods."

'But would poor people really have to make such a sacrifice to eat healthily? The report implies this, but its weakness is that it doesn't investigate it properly. The challenge now is to find out whether a typical social class D and E diet can be adapted to be made healthy and yet still both affordable – and acceptable – to the poor. The report suggests not. Its solution is to give the poor more money. But where is the evidence to suggest any extra money would be spent on food?

'Tightening Belts' is published @ £5 by the London Food Commission, 88 Old Street, London EC1.

	Typical low-income meals for four people	Cost	Typical NACNE meals for four people	Cost
BREAKFAST:	tea		tea	
	4 biscuits	10p	12oz muesli	43p
			1pt skimmed milk	23p
LUNCH:	$1\frac{1}{2}$lb sausages	£1.36$\frac{1}{2}$	$1\frac{1}{2}$lb mackerel	£1.27$\frac{1}{2}$
	$1\frac{1}{2}$lb potatoes	50p	$1\frac{1}{2}$ potatoes	50p
	oil (for chips)	3$\frac{1}{2}$p	1lb carrots	30p
	12oz ice cream	35p	500ml plain yoghurt	72p
	can of pears	51p	4 apples	50p
	Total	£2.86	Total	£3.96

Total calories per person for breakfast and lunch in both low-income and NACNE examples is 959: if the calories aren't the same, a comparison isn't valid. Notice mackerel costs less per pound than sausages, yet the healthy meals can still work out more expensive. Examples by Isobel Cole-Hamilton.

13 Review the various comparisons made in the following three tables and write briefly about the connections among the data in each table and any conclusions and any questions you have concerning the given information.

Comparison of median earnings of full-time workers by sex (persons fourteen years and over)

Year	Median earnings ($)	
	Women	Men
1960	3,293	5,417
1965	3,823	6,375
1970	5,323	8,966
1971	5,593	9,399
1972	5,903	10,202
1973	6,335	11,186
1974	6,772	11,835
1975	7,504	12,758
1976	8,099	13,455

Source: US Department of Commerce, Bureau of the Census

Median incomes of full-time workers by occupation (persons fourteen years and over)

Major occupation group	1976 income ($)	
	Women	Men
Professional and technical workers	11,081	16,296
Non-farm managers and administrators	10,177	17,249
Sales workers	6,350	14,432
Clerical workers	8,138	12,716
Operatives (including transport)	6,696	11,747
Service workers (except private household)	5,969	10,117

Source: US Department of Labor, Women's Bureau

Median income comparisons of full-time workers by educational attainment, 1976 (persons twenty-five years and over) ($)

Years of school completed	Women	Men
Elementary school:		
Less than 8 years	5,644	8,991
8 years	6,433	11,312
High school:		
1 to 3 years	6,800	12,301
4 years	8,377	14,295
College:		
1 to 3 years	9,475	15,514
4 years or more	12,109	19,338

14 The following 'Dear Abby' (*Boston Herald American*, 6 October 1981) was brought to my attention by one of my students, Darleen Bonislawski.

'Dear Abby:

'In a recent column in the *Delaware State News*, a faithful girl Friday, signing herself "Lakewood, Calif.", said that her boss would give her $200 if she could guess within $200 the price of an elegant sofa shown in a magazine.

'"Lakewood" said she guessed $2,800. The boss said the price was $3,000 and refused to give her the $200, saying her guess was just $1 low.

'You concurred with her calculation that $2,800 is within $200 of the $3,000, but I think she missed it by a whisker – Abe Lincoln's whisker on a penny, that is.

'To be within $200 of $3000, her guess would have had to be at least $2800.01. The boss was wrong too. Her guess was 1 cent low instead of $1.

'Picayune Penny in Dover, Del.'

What is your opinion?

Recreation

15 What mathematical symbol can be placed between a 4 and 5 to make a number greater than 4 and less than 5?

16 What is the relationship of the artwork by Karen Shaw on the part-title (page 73) to this section?

REVIEW QUIZ (comparing the size of whole numbers and decimal fractions) (*Answers at end of chapter*)

1 Replace the question mark with the correct comparison symbol:
(a) 3040 ? 3404 (b) 3.04 ? 3.004

2 According to the Boston *Globe* (30 March 1979), the five highest average salaries of professional major league baseball teams are:

Boston Red Sox	$147,803
California Angels	$141,814
Los Angeles Dodgers	$135,884
New York Yankees	$188,880
Philadelphia Phillies	$159,039

(a) Which team had the highest average salary?
(b) Which team had the lowest average salary?

3 Use the following table to determine in which year UK direct investment in South Africa was the greatest:

UK net direct investment and earnings in South Africa, 1981–((in £ million)

Year	Net investment	Net earnings
1981	291.6	465
1982	194.4	411
1983	296.0	527
1984	120.3	393
1985	199.0	362
1986	86.0	320

Source: *British Business*, 3 July 1987 and 11 March 1988

ANSWERS: CHAPTER 12

Solved problems ('You try these')

1 (a) 65,438 < 65,483 (b) 125,126 < 126,125 (c) 12,345 > 2,345

2 6.784 < 6.847

3 7 < 7.2

4 9 > .95

5 .52 = .520

6 35,325 > 35,235 > 352 > 325 > 35.25 > 35.2 > 0.52 > .352

7 Morris, Illinois

Unsolved problems

1 (a) 3,352 > 3,252 (b) 46,784 < 49,784 (c) 2,475,235 < 24,752,350 (d) 8.65 > 8 (e) 9 > .98 (f) 1 < 1.002 (g) .349 < 1 (h) .304 = .3040

2 (a) 48 < 84 < 804 < 8,004
(b) 1,342 < 12,342 < 12,432 < 13,423

3 (a) 15.53 > 15 > 13.2 > 13.04
(b) 0.303 > 0.30 > 0.030 > 0.003 > 0.0003 > 0.00003

4 All the cartoons' humour is based on a comparison of amounts: in the 'pollutant spill' one, the gigantic amount of polluted waste is compared to the tiny bucket for the super-fund clean-up; in 'the history of the arms reduction talks', the comparison is from frame to frame, comparing the arms *increase* by each side as the talks continue, so that since both sides have so many arms a comparison between the sides becomes meaningless; in the 'nuclear freeze' cartoon, the humour is the illogic of Reagan's comparison argument that if we *increase* arms, then we can *reduce* them, and still have more than we started with.

8 Johns Hopkins University, Massachusetts Institute of Technology, University of California, Illinois Institute of Technology, Stanford University, University of Texas, University of Rochester, Georgia Technical Research Institute, University of Dayton, Pennsylvania State University, University of Southern California, University of Washington, University of Alaska, Carnegie Mellon University, University of Illinois, University of New Mexico, California Institute of Technology, Harvard University, University of Pennsylvania, Columbia University.

9 Walking 2 mph, gardening, slow bicycling, slow dancing, walking 3 mph, swimming 20 yds/min, walking 4 mph, moderate dancing, moderate tennis, moderate basketball, moderate bicycling, swimming 45 yds/min, vigorous tennis, walking 5 mph and fast dancing, vigorous basketball, swimming 50 yds/min, fast bicycling, running 6 mph, running 10 mph, running 12 mph.

10 The chart shows a great increase in the number of students, schools, and teachers since the revolution in 1960. (It also shows more careful records being kept – less unavailable data.) Some other kinds of related comparisons for which I have data are as follows: About half of Cuba's children were in school in 1956; virtually all children ages six to twelve go to school as of 1981. More than 82 percent of the thirteen- to sixteen-year-olds are enrolled in secondary school, a figure thirteen times higher than that for 1958–9. The 1959 education budget was 12 pesos per person, and the 1980 budget was 137 pesos per person (Leiner, p. 205).

11 First of all, the comparisons suggested by these data can be greatly clarified by ordering them in a tabular form:

US corporations' investments and profits in various areas outside the USA, 1950–65

	Investment	Profits
Europe	$8.1 billion	$ 5.5 billion
Latin America	3.8 billion	11.3 billion
Africa	5.2 billion	14.3 billion

Now, it is clear that although US corporations invested much less in Latin America and Africa than in Europe, they made much larger profits in Latin America and Africa than in Europe. (A probable explanation is much lower wages paid in Latin America and Africa.) Note that the 'Profits' category means additional money obtained *after* expenses, wages, and so on are paid. They didn't lose money in Europe, for example, even though the profit was smaller than the investment.

13 The first chart shows women's median income remaining considerably smaller than men's over the years 1960–1976. (After we learn about percentages in chapter 8, you'll see that finding the percentage women's income was of men's income clarify whether the comparative gap has increased or decreased.) The next two charts, using data for 1976, answer questions that might be raised concerning the data in the first chart. The occupation chart shows that the reason for the income gap is not simply that women are in low-paying jobs (like secretaries) because, even in the same field, women's income is significantly less than men's income. The educational attainment chart shows that the income gap cannot be explained by women, even in the same occupation, having less education that men, because with the same educational attainment women still have significantly smaller median incomes than me. (Note: even if the wage gap were due to women being in lower-paying jobs or having less education than men, one could question why that was the case.)

14 The key here is that the English phrase 'within' is not defined precisely mathematically. If 'within' means 'strictly less than', then the boss is right (although, yes she is off by 1 penny, not $1). But if 'within' means 'less than or equal to' then she is right.

15 A decimal point! $4 < 4.5 < 5$.

16 In Karen Shaw's system (described in Chapter 2, problem 11), the letters in 'less' add up to 55, and the letters in 'more' add up to 51. So, since when we compare 55 and 51, $55 > 51$, we can say 'less' is 'more', meaning that the sum of the letters of 'less' is more than the sum of the letters of 'more'. This is also a famous saying by architect Mies Van der Rohe. What else does it mean to you?

Review quiz

1 (a) $3040 < 3404$ (b) $3.04 > 3.004$

2 (a) New York Yankees (b) Los Angeles Dodgers

3 1983

13 Rounding whole numbers and decimals

According to Britain's Overseas Development Administration, British aid to South America between 1970 and 1976 was less than £20 million. Compare that figure to those in the following table.

Net earnings of UK companies in South America between 1970 and 1976

Year	£ million
1970	27.0
1971	23.3
1972	35.5
1973	54.7
1974	59.2
1975	81.7
1976	109.4
1970–6 total	£390.8 millions in 7 years

Source: Department of Industry, Business Statistics Office, HMSO, 1978

You might conclude that British companies earn far more money in South America than the British government gives in aid to South America. For example, in 1976 alone British companies earned about £110 million, more than five times the total foreign aid in all the years 1970–6. Even though the number on the chart is £109.4, in most real-life situations we do not use an 'exact' figure. A number sensibly close to the exact figure gives a reasonably accurate picture of the situation. In addition, in our very complex, giant world it is virtually impossible to get the exact number. In this case, companies may report their earnings differently, making it unclear whether to count certain items as earnings or expenses.

When we say that 109.4 million is approximately equal to 110 million, we are rounding the 109.4 to the nearest ten. (Of course, if we consider the 'billions' we are rounding to the nearest ten billion.) When we round 109.4 to the nearest ten what we are actually doing is:

1. Counting by tens.
2. Locating the two tens (100 and 110) between which 109.4 lies.
3. Determining whether 109.4 is closer to 100 or to 110. Since 105 is half-way between 100 and 110, and 109.4 > 105, we conclude that 109.4 is closer to 110 than to 100.

A number line picture shows this clearly:

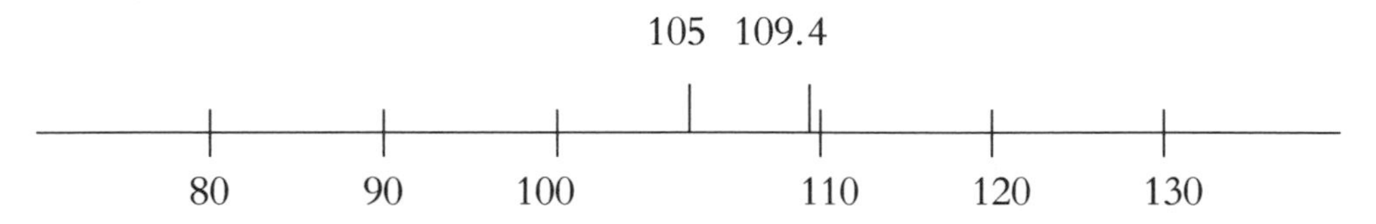

The short cut for rounding a number to a given place-value is:

1. Locate the place to which you are rounding.
2. Examine the place-value to the immediate right of the place to which you are rounding (that is, the next smallest place-value).
3. (a) If the digit in that place is less than 5, replace it and all digits to the right of it with zeros.
   (b) If the digit in that place is 5 or more, replace it and all digits to the right of it with zeros *and* add 1 to the digit in the place to which you are rounding.

Remember that we can drop zeros at the end of a number with a decimal point. Therefore, some of the zeros created by rounding may be dropped.

In this example, since we are rounding 109.4 to the nearest ten, we examine the 9 in the units' place. 9 > 5, so we change the 0 in the tens' place to 1 and replace the 9 and the 4 with zeros: 110.0. Since we can drop zeros at the end of a number with a decimal point, and we can also drop the decimal point if the number is a whole number, we say 109.4 rounded to the nearest ten is 110. Of course, you often use this rule informally. I state it here because some of the numbers are more difficult to round informally and because understanding rules for basic math forms a foundation for understanding how numbers work in complex situations and for further developing your mathematical intuition.

SOLVED PROBLEMS

(*Answers below*)

Drill

Problem 1 Round 752 to the nearest hundred.

Solution Because we are rounding to the nearest hundred we examine the digit in the tens' place. That digit is 5, so we add 1 to the 7 and replace the other digits with zeros.

Answer 800 (we write 752 ≑ 800 where ≑ means 'is approximately equal to')

Note The above method is a short cut for locating 752 between 700 and 800 and determining that 752 is closer to 800.

You try this (*Answer at end of chapter*)

Round 572 to the nearest hundred.

Problem 2 Round 234,000 to the nearest ten thousand.

Solution Although this number is rounded in the sense that it has zeros at the end, it is not rounded to the nearest ten thousand. (If the problem had been to round 234,000 to the nearest thousand, the answer would have been 234,000.) Because the digit in the thousands' place is smaller than 5, we round 234,000 to

Answer 230,000

You try this (*Answer at end of chapter*)

Round 463,900 to the nearest ten thousand.

Problem 3 Round 129,999 to the nearest thousand.

Solution The digit in the hundreds' place is greater than 5, so we add 1 to the digit in the thousands' place and replace the digits to the right of the thousands place with zeros. However, because we are adding 1 to a 9 in the thousands' place, we get 10 thousands, or a zero in the thousands' place and one more in the ten thousands' place.

Answer 130,000

Note Following rules can get complicated, as in this case. Never lose sight of what the numbers mean. 129,999 is only 1 short of 130,000 so our answer certainly makes sense. In cases like this, the number rounded to the nearest thousand is the same as the number rounded to the nearest ten thousand.

You try this *(Answer at end of chapter)*
Round 379,999 to the nearest thousand.

Problem 4 Round 302 to the nearest ten.

Solution The number in the ones' place is less than 5, so we replace it with a zero.

Answer 300

Note This answer is the same whether the number is rounded to the nearest ten or to the nearest hundred.

You try this *(Answer at end of chapter)*
Round 603 to the nearest ten.

Problem 5 Round 34.048 to the nearest hundredth.

Solution Because we are rounding 34.048 to the nearest hundredth, we examine the digit in the thousandths' place. Since $8 > 5$, we add 1 to the 4 in the hundredths' place and replace the digits to the right of the hundredths' place with zeros. 34.048 rounds to 34.050. Since $34.050 = 34.05$, we usually say 34.048 rounded to the nearest hundredth is

Answer 34.05.

You try this *(Answer at end of chapter)*
Round 65.067 to the nearest hundredth.

Problem 6 Round 15.96 to the nearest tenth.

Solution Since the digit in the hundredths' place is larger than 5, we drop it and add 1 to the digit in the tenths' place. However, because we are adding 1 tenth to 9 tenths, we get 10 tenths or 1 whole.

Answer 16.0 or 16

Note We dropped the digit in the hundredths' place instead of replacing it with a zero because zeros at the end of the number written with a decimal point do not change the value of the number. $16.00 = 16.0 = 16$.

You try this *(Answer at end of chapter)*
Round 17.97 to the nearest tenth.

Applications

Problem 7 Using the table at the beginning of this chapter, find the approximate net earnings of UK companies in South America in 1975.

Solution In real life, you are often expected to use your judgement about how precise an estimate is needed. The more experience you have working with numbers, the more confidence you will gain in your mathematical judgement. You control numbers; they do not control you. In this case they do not contain some hidden, trick message telling how much to round them off. In real life, your reason for wanting to know the net earnings would determine whether you rounded £81.7 million to £82 million or £80 million, or even whether you felt £81,700,000 was an approximate enough answer. Also, if you wanted to emphasize how large an amount this was, you could use the fact that, rounded to the nearest hundred million, £80 million is close to £100 million and £100 million = £.1 billion or $\frac{1}{10}$ of a billion pounds.

Note The fact that this decision is yours may seem intimidating now. After all, you may feel it's hard enough to round when told what place to round to. As you get more comfortable using numbers, the fact that you can use your common sense in maths will I hope give you added self-confidence.

You try this (*Answer at end of chapter*)

Find the approximate net earnings of UK companies in South America in 1971.

Problem 8 Using the following chart (*New York Times*, 20 November 1977), find the average salary for a woman, to the nearest thousand dollars, in 1969 in the USA.

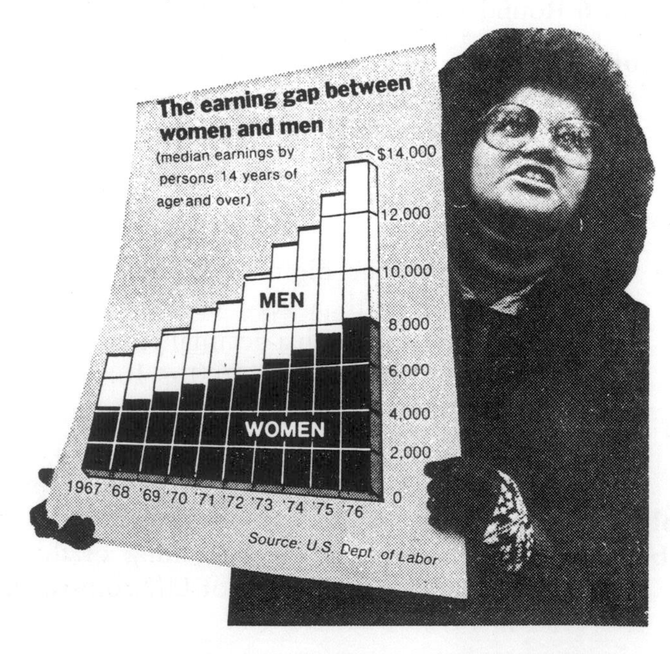

Solution This numerical information is presented in the form of a *bar graph*. Each bar represents the average salary for women and men in the year written at the base of the bar. The range of average salaries (from $0 to $14,000) is written along the vertical side. The total height of each bar indicates men's average salary; the height of the darkened part of each bar represents women's average salary. We read a bar by 'drawing' an imaginary horizontal line from its top to the vertical scale.

In this case, we're interested in the height of the darkened part of the 1969 bar. Notice that there are lines drawn across the bars for each of the vertically listed salaries. Since the height we are measuring is almost half-way between the $4,000 and $6,000 mark, it represents an average salary of about $5,000.

Answer In 1969, the average salary of a woman was $5,000.

Notes
1 Some bar graphs write the exact figure on each bar. When this isn't done, as in this case, you have to estimate.
2 One of the reasons data are presented in bar graphs is because the general trends in the information can be seen at a glance. We don't have to compare numbers to see that over the years women's average salary has remained about half of men's average salary, even though both average salaries have risen.

You try this (*Answer at end of chapter*)

Find the average salary of a man in 1969 to the nearest thousand dollars.

Problem 9 Using the information contained in the bar graph from problem 8, create a maths problem whose solution involves rounding numbers.

Solution Creating problems is very difficult. Here, you must not only understand how to round numbers and read bar graphs, but you also must know which questions can and which questions cannot be answered from the given information. For example, using this chart you cannot find the average salary of men in 1955, and you cannot find the average salary of black men in 1970. You can find the average salary of men and women in any of the years from 1967 to and including 1976. If you understand subtraction, you can also find the gap, or difference, between men's and women's average salaries for each of the years from 1967 through 1976. Also, we can then see if the gap is getting smaller or larger.

Answer Find the average salary of a man in 1972.

You try this

Create another problem whose solution involves rounding numbers.

UNSOLVED PROBLEMS

(Answers at end of chapter)

Drill

1 Round to the nearest hundred:
(a) 348 (b) 6,348 (c) 762 (d) 12,164

2 Round to the nearest ten:
(a) 15,672 (b) 308

3 Round to the nearest million:
(a) 1,009,465 (c) 342,000,000
(b) 15,925,324 (d) 3,420,000

4 Round to the nearest hundred thousand:
(a) 2,095,468 (b) 13,025,481

5 Round to the nearest thousand:
(a) 2,999 (c) 42,345,542
(b) 9,999 (d) 42,345,042

6 Round to the nearest ten thousand:
(a) 15,820 (c) 2,034,250
(b) 11,820 (d) 2,000,125

7 Round to the nearest hundredth:
(a) 15.274 (c) 462.672 (e) 315.074
(b) 5.265 (d) 3.604 (f) 0.006

8 Round to the nearest whole number:
(a) 14.7 (c) 5.96 (e) 436.5 (g) 100.9
(b) 14.07 (d) 5.39 (f) 439.6 (h) 109.8

9 Round to the nearest thousandth:
(a) 14.0235 (c) 0.0034 (e) 16.0007
(b) 14.0231 (d) 0.0015 (f) 9.9995

10 Round to the nearest tenth:
(a) 19.84 (c) 123.446 (e) 12.008
(b) 123.456 (d) 12.07 (f) 0.08

11 For a challenge, round to the nearest millionth:
(a) 0.0000023 (b) 0.0000509

Thought

12 Find a real-life situation in which the rules for rounding are not followed.

13 (a) Create and solve a quiz which reviews the major concepts of this chapter. Include at least three applications using real-life data.
(b) Write briefly about which of your questions is the most difficult and why; which question is easiest and why.

Applications

14 Use the following table (from Stone) to solve this problem:

SAT* average score	Family's median income ($)
750–800	22,425
700–749	21,099
650–699	19,961
600–649	18,906
550–599	17,939
500–549	16,990
450–499	16,139
400–449	15,240
350–399	14,068
300–349	12,384
250–299	9,865
200–249	7,759

* Scholastic Aptitude Test, on which for admission many US colleges require high scores

(a) What conclusions might you draw from the information in this table?
(b) Create and solve three problems using the information in the table whose solutions involve rounding.

15 Study the following table from the *1979 Information Please Almanac* (Information Please Publishing Inc., 1978):

United States audience composition of selected prime time programme types (in millions)*

Audience	General drama	Suspense and mystery drama	Situation comedy	Adventure	Variety
Women (18 years) and over)	10.71	10.58	11.83	10.22	9.48
Men (18 and over)	6.86	8.37	8.74	9.07	7.38
Teens (12–17)	2.25	2.68	3.72	3.01	2.74
Children (2–11)	2.77	3.00	4.55	6.42	3.73

* All figures are estimates for the period October–December 1977.

Source: A.C. Nielsen Company, Nielsen Television Index Audience Estimates

(a) Summarize the kind of information contained in this table.
(b) What conclusions might you draw from the information in this table?
(c) Which type of programme do the most women (eighteen years and over) watch?
(d) To the nearest million, estimate how many women watch that programme type.
(e) About how many men (eighteen years and over) watch adventure shows?
(f) Create and answer a maths problem, based on the information in this table, whose solution involves rounding numbers.

16 Discuss the general point of the following bar chart. Use the information to create and solve two maths problems whose solution involves rounding numbers.

Income distribution in Brazil and the UK: share of gross national product

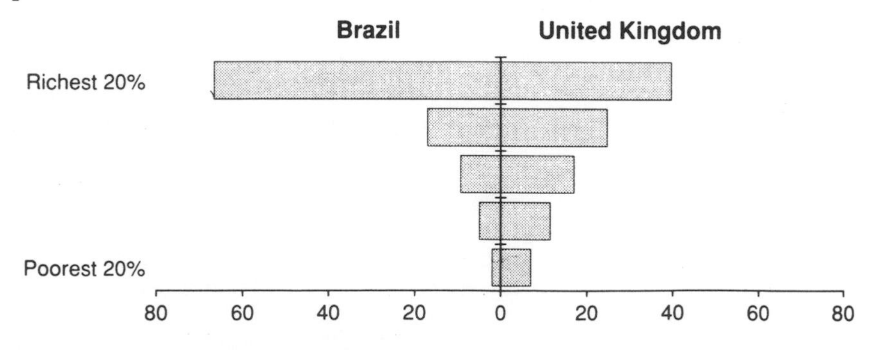

Source: *World Development Report*, 1987

17 Use the following line graph from the *1977 Statistical Abstract of the United States*, to solve this problem.

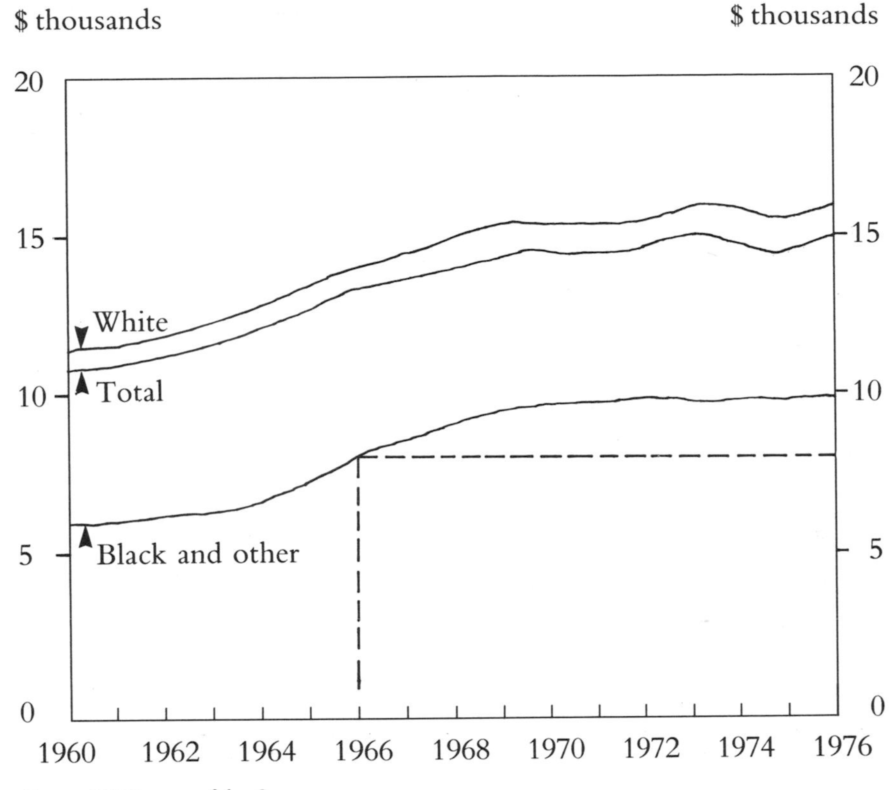

Source: US Bureau of the Census

Note: Line graphs are read in a similar way to bar graphs, except you have to 'draw' two imaginary lines. First locate the year you're interested in and 'draw' an imaginary vertical line to the graph line you're studying. Then, to read off the salary, 'draw' an imaginary horizontal line to the salary scale. For example, in 1966, the average income of black and other families of colour was about $8,000 (somewhat more than half-way between $5,000 and $10,000).

(a) What general conclusions can you draw from the information on this line graph?
(b) Find the average salary for a US family in 1973.
(c) Find the average salary for a US white family in 1973.
(d) Find the average salary for a US black family and family of colour in 1973.
(e) Create and answer a maths problem based on the information on this line graph.

18 Summarize the main point of the following charts (*New Society Database*, 23 January 1987). Use the information to create and solve three problems whose solutions involve rounding numbers.

Average weekly wages in Britain

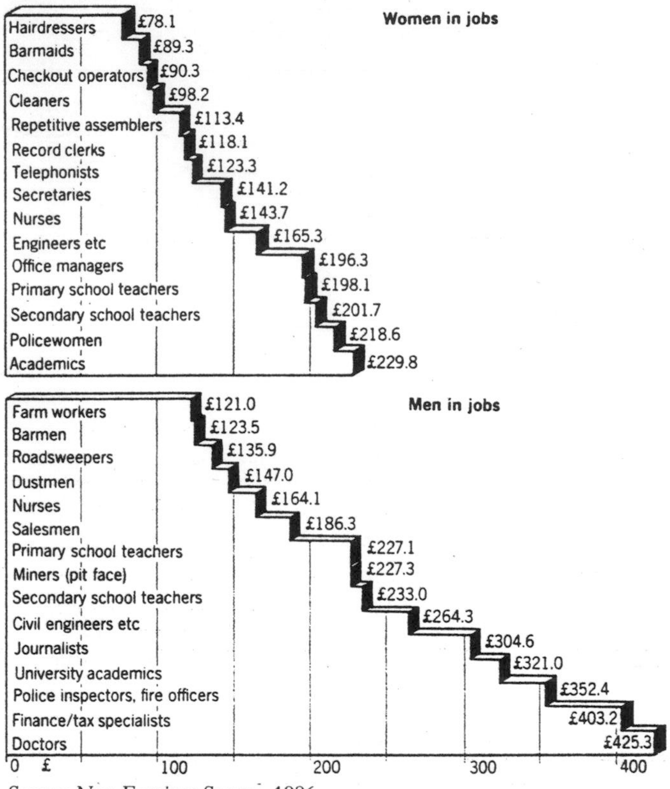

Source: New Earnings Survey, 1986

19 As I discussed in the Introduction, unemployment statistics vary depending on who is counted as unemployed. A more comprehensive concept, underemployment, includes:

- unemployed workers looking for work but unable to find a job;
- discouraged workers, who are unemployed and want work but are not actually looking for a job because they believe no jobs are available;
- involuntary part-time workers who want full-time work but are unable to find it; and
- workers in jobs which provide income inadequate to support a decent standard of living.

The following tables from the USA (Gordon, pp. 73–4) include low, middle, and high estimates, which vary according to estimates of the 'hidden' underemployed, who are not found by the census-takers, in urban ghettos, rural areas, and prisons. Use the tables to create and solve five problems whose solutions involve rounding numbers.

Counting the underemployed, 1975 (all numbers in thousands)

	Low estimate		Middle estimate		High estimate	
	Number	%	Number	%	Number	%
Expanded labour force	93,695	100.0	95,232	100.0	97,809	100.0
1 Unemployed	7,830	8.4	7,830	8.2	7,830	8.0
2 Discouraged workers	1,082	1.2	2,619	2.7	5,196	5.3
3 Involuntary part-time	3,748	4.0	3,748	3.9	3,748	3.8
4 The working poor	3,234	3.5	7,146	7.5	15,274	15.6
Total underemployment	15,894	17.0	21,343	22.4	32,048	32.8

Note: Totals do not add to total percentages owing to roundings.

Some find it hardest: underemployment among blacks and women, 1975 (all numbers in thousands)

	Low estimate		Middle estimate		High estimate	
	Number	%	Number	%	Number	%
Blacks						
Expanded labour force	10,835	100.0	11,295	100.0	12,049	100.0
1 Unemployed	1,459	13.4	1,459	12.9	1,459	12.1
2 Discouraged workers	306	2.8	766	6.8	1,520	12.6
3 Involuntary part-time	638	5.9	638	5.6	638	5.3
4 The working poor	615	5.7	1,266	11.2	2,523	21.0
Total underemployment	3,018	27.8	4,129	36.6	6,140	51.0

Note: Numbers are for 'Negroes and other races'.

	Number	%	Number	%	Number	%
Women						
Expanded labour force	37,720	100.0	38,746	100.0	40,457	100.0
1 Unemployed	3,445	9.1	3,445	8.9	3,445	8.5
2 Discouraged workers*	722	1.9	1,748	4.5	3,459	8.5
3 Involuntary part-time†	1,855	4.9	1,855	4.8	1,855	4.6
4 The working poor	635	1.7	2,702‡	7.0	7,964	19.7
Total underemployment	6,657	17.7	9,750	25.2	16,723	41.3

Note: Totals do not add to total percentages owing to rounding.

* This set of figures involves some approximation through interpolation.

† This figure involves a minor adjustment for agricultural workers.

‡ This is the one case where the two alternative 'middle' estimates are significantly different from each other; this number, therefore, has an elusive substantive meaning.

20 Use the article below (from *The Progressive*, November 1981) to answer the following questions:

(a) Read this article and discuss the author's main point, tone, and your reactions to his point.

(b) Why did Bush's Press Secretary use the figure $125,000 when the actual amount raised was $124,700?

(c) About how much was allocated for repairs and new Vice-Presidential furnishings in the 1982 Federal budget?

(d) Create and solve a maths problem based on the information in this article whose solution involves rounding numbers.

(e) Guess how much money the White House furnishings are worth.

(f) Discuss how you might find out the worth of the White House furnishings.

'Old values, anyone?

'Shortly after word leaked out that George and Barbara Bush had raised $124,700 in private, tax-deductible donations to redecorate the Vice-President's official residence, they called off the fund drive. "Mrs Bush finds $125,000 a sufficient amount to refurbish the rooms," said the Vice-President's press secretary, Peter Teeley. Fortunately, there will also be the $51,700 for repairs and new Vice-Presidential furnishings proposed in the Federal budget for fiscal 1982.

'Over at the White House, meanwhile, the Reagans have laid out $209,508 for a new set of Presidential chinaware. That money came from another privately raised tax-deductible fund – $822,641 in all, much of it contributed by oil entrepreneurs.

'What's wrong with all this? Not much, really. Who would want the Vice-President of the United States to sleep in a sagging bed under a leaky roof? Who would want the President to gobble his jellybeans off chipped five-and-dime dishes? Surely the donors to the "private" Presidential and Vice-Presidential residential funds are not buying influence: would Reagan and Bush be hostile to the oil industry if it weren't for the contributions? And if there is a tax loss to the Treasury, it certainly is small potatoes compared to all the other dodges and loopholes.

'It's just that there's a *principle* involved here. We hate to see able-bodied Americans – folks perfectly capable of fending for themselves – relying on handouts from government and private charity. It makes you wonder what the country's coming to.'

REVIEW QUIZ (rounding whole numbers and decimals)

(*Answers at end of chapter*)

1 Round to the nearest thousand: 40,999.

2 Round to the nearest hundredth: 15.472.

3 Round to the nearest million: 346,782,000.

4 Use the audience composition table earlier in this chapter (problem 15) to find the number of children (age two to eleven) who watch adventure shows, rounded to the nearest hundred thousand.

5 Use the line graph in this chapter on US families' average income to estimate the average salary for black families and families of colour in 1969.

ANSWERS: CHAPTER 13

Solved problems ('You try these')

1 600
2 460,000
3 380,000
4 600
5 65.07
6 18.0 or 18
7 £20 million or (since 23.3 is so close to the midpoint between £20 and £30, we sometimes 'disobey' the rules and round this off to) £25 million
8 $8,000

Unsolved problems

1	(a) 300	(b) 6,300	(c) 800	(d) 12,200
2	(a) 15,670	(b) 310		
3	(a) 1,000,000			
	(b) 16,000,000			
	(c) 342,000,000 (this number is already rounded to the nearest million)			
	(d) 3,000,000			
4	(a) 2,100,000	(b) 13,000,000		
5	(a) 3000	(b) 10,000	(c) 42,346,000	(d) 42,345,000
6	(a) 20,000	(b) 10,000	(c) 2,030,000	(d) 2,000,000
7	(a) 15.27	(b) 5.27	(c) 462.67	(d) 3.60 or 3.6
	(e) 315.07	(f) 0.01		
8	(a) 15	(b) 14	(c) 6	(d) 5
	(e) 437	(f) 440	(g) 101	(h) 110
9	(a) 14.024	(b) 14.023	(c) 0.003	(d) 0.002
	(e) 16.001	(f) 10.000 or 10		
10	(a) 19.8	(b) 123.5	(c) 123.4	(d) 12.1
	(e) 12.0 or 12	(f) 0.1		
11	(a) 0.000002	(b) 0.000051		

12 One example is when you buy an item in the supermarket that sells three for a dollar (or 33.3 cents per item): the supermarket rounds the price up to 34 cents, whereas the maths rule would round 33.3 to 33.

14(a) The higher median income, the higher SAT score. Why do you think this occurs?

15(a) The chart breaks down the viewing audience into children, teens, men adults and women adults. For each person category, the chart lists how many millions watch the various types of TV shows, broken into five categories of shows.

(b) One striking observation, based also on the fact that there are approximately equal numbers of men and women adults, is how much more TV women watch than men (maybe because women are home more?). Also interesting is how children strongly favour adventure shows.

(c) Situation comedy.

(d) 12 million.

(e) I would round this off to 9 million, but it wouldn't be wrong to round to 9.1 million or 9,100,000.

16 The chart shows that the distribution of gross national product (GNP) is extremely unequal in both the UK and Brazil, and even more so in Brazil. For example, in the UK the richest 20 percent of households receive about 40 percent of the GNP; in

Brazil the richest 20 percent receive about 70 percent of the GNP! If the GNP were equally distributed, 20 percent of the population would get 20 percent of the GNP.

17(a) The average income of black families (and families of colour) has continued to be significantly lower than white families' income over the years from 1960 to 1976. This was still the case at the time when this book was largely written (1986–7).

(b) $15,000

(c) $16,000

(d) $10,000

18 According to *New Society* magazine (23 January 1987),

'Even when men and women are in the same occupation, they can be getting very different pay. A male primary teacher gets £227.10 a week, on average; a woman £198.10. Although rates for the two sexes are the same in teaching, males tend to fill posts higher up the pay scales, or to have longer service, thus accumulating more increments.

'The difference in pay between the sexes is just one example of inequality. As the charts show, equally striking are the divergences between different occupations, with doctors, for example, earning roughly four times as much as farm labourers.'

As the magazine also notes,

'certain highly paid occupations are virtually confined to men. In compiling the *New Earnings Survey*, the statisticians failed to come up with enough female doctors or journalists to calculate an average. Meanwhile, low-paid occupations, such as hairdressing, are dominated by women.'

20(a) The author sarcastically questions the large amount of 'welfare' that Bush and Reagan have used for furnishings and china (according to the October 1981 issue of *Dollars & Sense*, the money for the china 'comes from tax-deductible contributions raised by the Reagans and the White House Historical Association to redecorate the White House. Assuming that the contributors are mostly in the top tax brackets, the taxpayers are in effect picking up 70 percent of the tab'!). He uses Reagan's arguments against real welfare to the poor ('we hate to see . . . folks perfectly capable of fending for themselves relying on handouts from government and private charity'), to compare the 'welfare-to-the-rich' to refurbish the White House to the 'welfare-to-the-poor' to feed kids . . .

(b) He rounded $124,700 to the nearest thousand dollars.

(c) $51,700 can be rounded to $52,000 (the nearest thousand) or $50,000 (the nearest ten thousand).

(e) Try to use the information in the article to make a reasonable guess (I don't know the answer . . .).

(f) One idea: a tourist brochure from the DC Chamber of Commerce might have the information or at least have the number of rooms with pictures from which you could use the numbers in the article to guess how much each room is worth . . .

Review quiz

1 41,000

2 15.47

3 347,000,000

4 6,400,000

5 $10,000

14 Different ways of writing fractions and mixed numbers

The following table compares the total number of men with the total number of women who have served in various branches of the US government over the past 200 years:

	Men	*Women*
Senate	1,715	11
House	9,591	87
Supreme Court	101	0
Cabinet	507	5

Source: New Jersey Women's Coalition Press, June 1978

Possibly a more dramatic picture could be painted by converting these data into fractions. For example, the exact fraction of women who have served in the Cabinet in $\frac{5}{512}$ ($512 = 507 + 5$ = total number of people who had served in the Cabinet up until June 1978). To get a better visual idea of what part of the total group of people who've served on the Cabinet were women, we can round off the fraction and write it in *reduced form*. We use the fact that if you divide the numerator and denominator of a fraction by the *same* number, the numerical quantity represented by the fraction remains the same.

$$\frac{5}{512} \doteq \frac{5}{500} \text{ and } \frac{5 \div 5}{500 \div 5} = \frac{1}{100}$$

So we can say that about $\frac{1}{100}$, or 1 out of 100, people who've served in the Cabinet have been women.

SOLVED PROBLEMS

(*Answers below*)

Drill

Problem 1 Write two different fractions which represent the shaded portion of the following diagram.

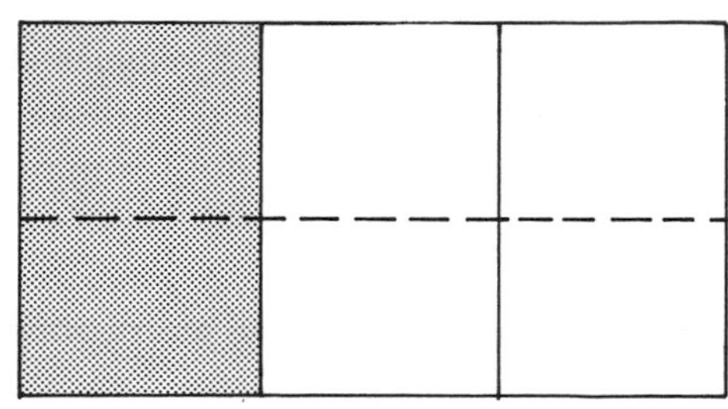

Solution We can consider the whole as being broken into six equal parts, two of which are shaded. Therefore $\frac{2}{6}$ of the whole is shaded. However, if we ignore the broken, horizontal line, then the whole is divided into three equal parts, one of which is shaded. So, the fraction $\frac{1}{3}$ also represents the shaded portion of the rectangle.

Answer The shaded portion of the diagram can be represented by $\frac{2}{6}$ or $\frac{1}{3}$. (Therefore, $\frac{2}{6} = \frac{1}{3}$.)

Note Dividing the numerator and denominator of $\frac{2}{6}$ by 2, we also get $\frac{2 \div 2}{6 \div 2} = \frac{1}{3}$. The diagram illustrates that this rule gives us reasonable results.

You try this (*Answer at end of chapter*)

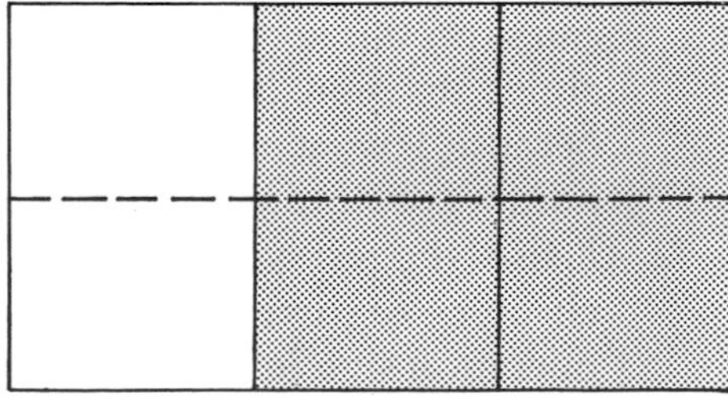

Problem 2 Write each of the following fractions in reduced form.

(a) $\frac{9}{21}$ (b) $\frac{9}{20}$ (c) $\frac{12}{30}$

Solution (a) In order to reduce a fraction you must find a number that can be divided without remainder into both the numerator and the denominator. This number is referred to as the *common divisor*. If the common divisor is not evident, start with any divisor of the numerator and check (by actually dividing) whether or not that divisor also goes into the denominator without remainder. In this case, we see that 3 is a common divisor.

$$\frac{9 \div 3}{21 \div 3} = \frac{3}{7}$$

Answer (a) $\frac{3}{7}$ is the reduced form of $\frac{9}{21}$.

Solution (b) In this case, there is no *common* divisor (other than 1), even though 9 can be divided without remainder by 9 and 3, and 20 can be divided without remainder by 20, 10, 2, 5, and 4.

Answer (b) $\frac{9}{20}$ is the reduced form.

Solution (c) In this case, you may see immediately that 6 is a common divisor.

$$\frac{12 \div 6 = 2}{30 \div 6 = 5}$$

However, if you first see that 2 is a common divisor, you will still eventually get the correct reduced form. First $\frac{12 \div 2}{30 \div 2} = \frac{6}{15}$. Then you see that 6 and 15 have a common divisor of 3, $\frac{6 \div 3}{15 \div 3} = \frac{2}{5}$. The important thing to remember is to make sure the numerator and denominator of the reduced fraction have no common divisors (other than 1).

Answer (c) $\frac{2}{5}$ is the reduced form of $\frac{12}{30}$.

Notes

1 These problems do assume you can divide by small numbers. If you can't, don't panic – your teacher, or someone in the group, can probably show you, or you can skim this section and return to it after you learn more about division in Volume 2. I present this topic here because I want to familiarize you with the meaning of numbers before you study the meaning of the operations.

2 I also suggest you start to use your calculator in this chapter: to divide, you punch the numbers and the operation sign in the calculator in the order in which they are written.

3 Also, there are short cuts for finding various divisors which you may want to remember:

(a) 2 is a divisor of any number whose units' digit is even (0, 2, 4, 6, 8). For example, 246 is divisible by 2, and 351 is not divisible by 2.

(b) 5 is a divisor of any number whose units' digit is 0 or 5. This rule would show you immediately that 135 and 260 are divisible by 5.

(c) 10 is a divisor of any number whose units' digit is 0. For example, 3,840 is divisible by 10, and 425 is not divisible by 10.

You try this *(Answers at end of chapter)*

(a) $\frac{14}{35}$ (b) $\frac{8}{45}$ (c) $\frac{16}{40}$

Problem 3 Write the following fraction in 'approximately reduced' form:

$$\frac{426}{789}$$

Solution As we saw in the introduction to this section, in real life, the fractions you form are usually not neatly reducible. So in order to get an idea of what part of the whole such a fraction represents, we round off the numerator and denominator first, and then reduce.

In this case: $\frac{426}{789} \doteq \frac{400}{800}$

and $\frac{400 \div 400}{800 \div 400} = \frac{1}{2}$

Answer $\frac{426}{789} \doteq \frac{1}{2}$

You try this *(Answer at end of chapter)*

$$\frac{316}{897}$$

Problem 4 Given that each circle represents one whole, write two expressions which represent the shaded portion of the following diagram.

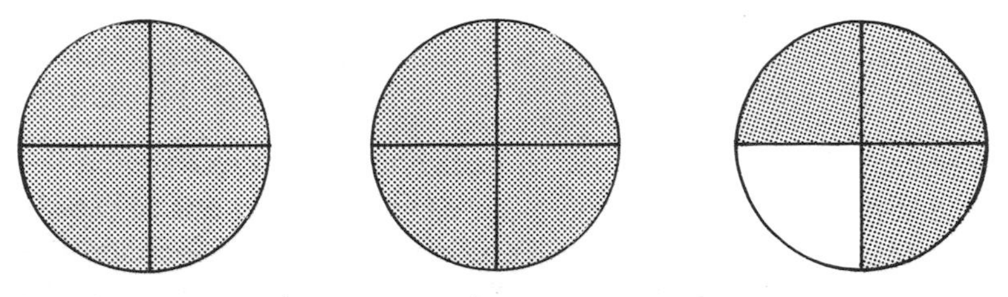

Solution Since two whole circles and $\frac{3}{4}$ of another circle are shaded, we can represent this diagram by $2\frac{3}{4}$ (read '2 and $\frac{3}{4}$'), which we call a *mixed number* because it represents an amount that is equal to a whole number plus a fraction.

Also, since there are eleven quarter-circles shaded, we can represent this diagram by $\frac{11}{4}$, which we call an *improper fraction* because it is a fraction that represents an amount greater than 1.

Answer The shaded portion of the diagram can be represented by $2\frac{3}{4}$ or $\frac{11}{4}$. (Therefore, $2\frac{3}{4} = \frac{11}{4}$.)

Note If you considered the entire diagram to represent one whole, then $\frac{11}{12}$ would represent the shaded portion because the whole diagram is divided into twelve equal parts, eleven of which are shaded. That is why the directions state that *each circle* represents one whole.

You try this (*Answers at end of chapter*)

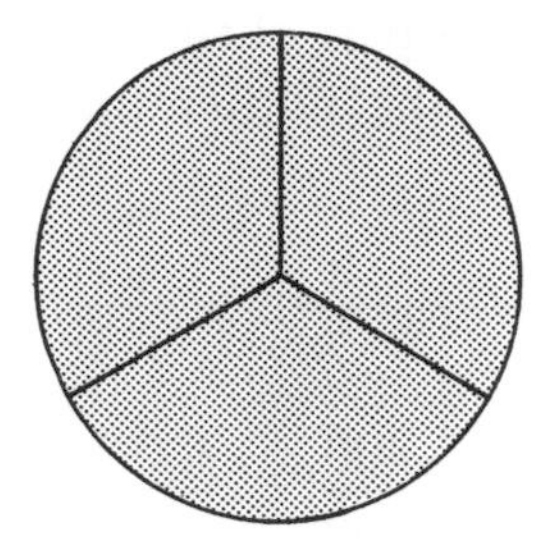 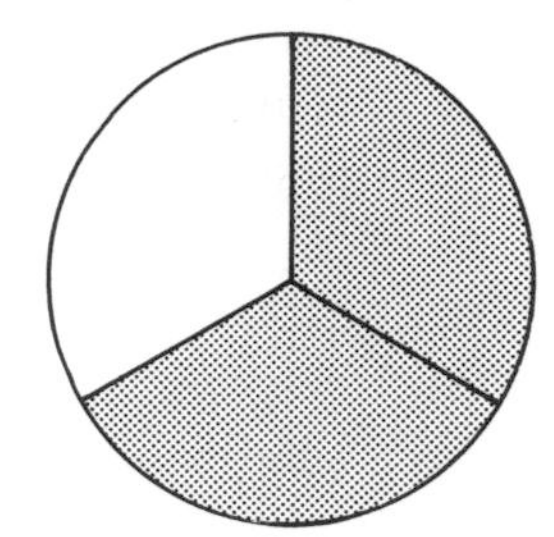

Problem 5 Write each of the following mixed numbers as improper fractions.

(a) $2\frac{3}{5}$
(b) $7\frac{2}{4}$
(c) 4

Solution (a) One way to solve this problem is to draw a diagram.

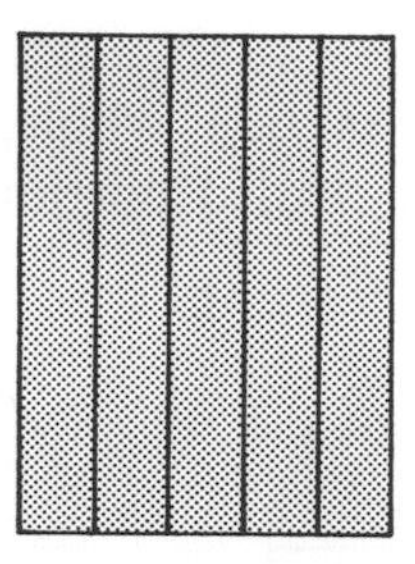 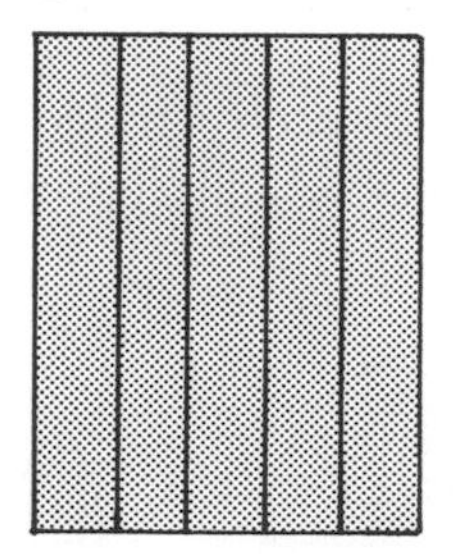 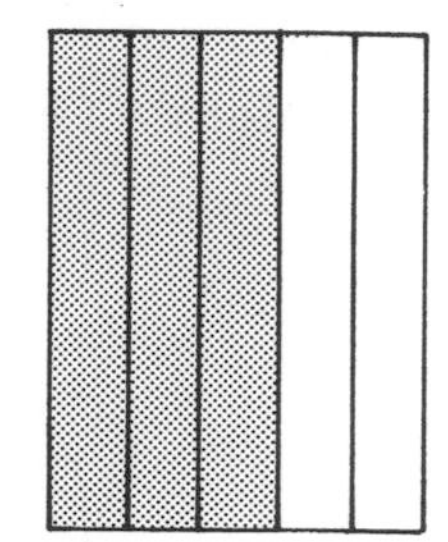

This diagram shows that $2\frac{3}{5} = \frac{13}{5}$. However, if the numbers were larger, the diagram method would be too time-consuming.

A short cut for changing a mixed number into an improper fraction is:

1 The numerator of the improper fraction is obtained by multiplying the whole number by the denominator of the

given fraction and adding the numerator of the given fraction.

2 The denominator of the improper fraction is the same as the denominator of the given fraction. In this case,

$$2\frac{3}{5} = \frac{(2 \times 5) + 3}{5} = \frac{10 + 3}{5} = \frac{13}{5}$$

Answer (a) $2\frac{3}{5} = \frac{13}{5}$

Notes

1 The rule makes sense. In this case, the given fraction tells you that wholes are being divided into fifths, so 2 wholes would contain $2 \times 5 = 10$ fifths. Since the mixed number represents 2 wholes (10 fifths) and 3 fifths, the mixed number also represents $10 + 3 = 13$ fifths.

2 You shouldn't really have to memorize this rule. If you forget it, draw a picture of a simple example, such as $2\frac{3}{4} = \frac{11}{4}$, ask yourself how you could get $\frac{11}{4}$ from $2\frac{3}{4}$, and that will probably stimulate your memory that

$$2\frac{3}{4} = \frac{(2 \times 4) + 3}{4} = \frac{11}{4}$$

3 These problems assume you can multiply (and add) small numbers. If you don't remember the basic multiplication facts, use your calculator to find the answers. We'll study addition and multiplication in detail in Volume 2.

Solution (b) Using the rule,

$$7\frac{2}{4} = \frac{(7 \times 4) + 2}{4} = \frac{28 + 2}{4} = \frac{30}{4}$$

Answer (b) $7\frac{2}{4} = \frac{30}{4}$.

Note Both the improper fraction and the fraction part of the mixed number can be reduced. If you need a reduced answer you could either reduce the improper fraction $\frac{30}{4}$ to $\frac{2}{15}$, or reduce $7\frac{2}{4}$ to $7\frac{1}{2}$ and then

$$7\frac{1}{2} = \frac{(7 \times 2) + 1}{2} = \frac{14 + 1}{2} = \frac{15}{2}$$

Solution (c) Since 4 is a whole number, not a mixed number, our rule does not apply (for example, there is no denominator of a given fraction by which to multiply 4). To represent a whole number as an improper fraction, we use the division meaning of fractions. Since $4 = 4 \div 1$, we can write $4 = \frac{4}{1}$.

Answer (c) $4 = \frac{4}{1}$

You try this *(Answers at end of chapter)*

(a) $2\frac{5}{6}$ (b) $9\frac{4}{12}$ (c) 6

Problem 6 Write each of the following improper fractions as mixed numbers.

(a) $\frac{13}{4}$ (b) $\frac{152}{6}$ (c) $\frac{20}{5}$

Solution (a) As in problem 5 (a), this problem can be solved by drawing a diagram, grouping every 4 fourths into 1 whole.

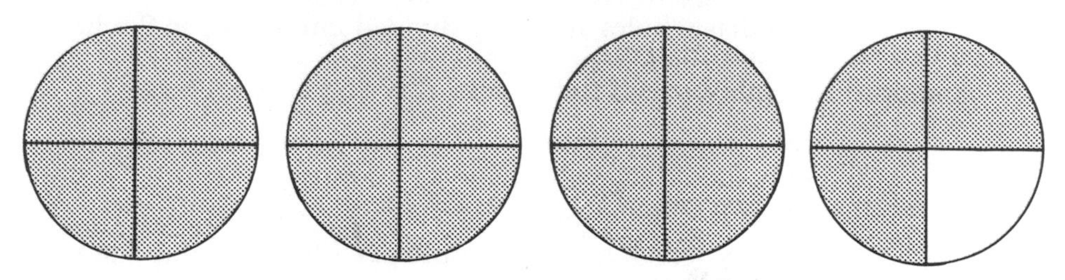

The diagram makes it clear that $\frac{13}{4} = 3\frac{1}{4}$.

However, you certainly don't want to spend time drawing a diagram for a problem such as $\frac{152}{6}$!

A short cut for changing an improper fraction into a mixed number is to use the division meaning of fractions:

1 Divide the denominator into the numerator to get the whole number.

2 The fraction part of the mixed number has a numerator which is the remainder of the division and has the same denominator as the improper fraction. In this case,

$$\frac{13}{4} = 13 \div 4 = 3 + \text{remainder } 1 = 3\frac{1}{4}.$$

Answer (a) $\frac{13}{4} = 3\frac{1}{4}$

Note This rule makes sense. Every group of 4 fourths is one whole, so since there are 3 such groups in 13, there are 3 wholes. The remainder 1 represents 1 of the 13 fourths, or 1 fourth.

Solution (b) $\frac{152}{6} = 152 \div 6 = 25 + \text{remainder } 2 = 25\frac{2}{6} = 25\frac{1}{3}$

Answer (b) $\frac{152}{6} = 25\frac{1}{3}$

Notes

1 $25\frac{2}{6}$ is also a correct answer. $25\frac{1}{3}$ is the reduced form of the answer since $\frac{2}{6} = \frac{1}{3}$.

2 You can check these problems by working backwards. Use the rule for changing mixed numbers to improper fractions to see that

$$25\frac{2}{6} = \frac{(25 \times 6) + 2}{6} = \frac{150 + 2}{6} = \frac{152}{6}$$

Solution (c) $\frac{20}{5} = 20 \div 5 = 4$. Since there is no remainder, this improper fraction is equal to a whole number.

Answer (c) $\frac{20}{5} = 4$.

Notes

1. Using the methods described above, you can see that $1 = \frac{1}{1} = \frac{6}{6} = \frac{15}{15} = \frac{362}{362}$, etc. Because any number divided by itself is equal to 1, 1 can be written as any fraction in which the numerator is equal to the denominator. In general, if n represents any whole number, then $1 = \frac{n}{n}$.
2. A 'proper' fraction, such as $\frac{7}{9}$, represents an amount smaller than 1 whole. In a proper fraction the numerator is always smaller than the denominator because proper fractions represent only a part of 1 whole. In $\frac{7}{9}$, for example, the whole is divided into 9 equal pieces. Seven of those 9 pieces are represented by the proper fraction $\frac{7}{9}$.
3. An 'improper' fraction, such as $\frac{13}{5}$, represents an amount greater than 1 whole. In an improper fraction the numerator is always greater than the denominator. The improper fraction $\frac{13}{5}$, for example, indicates that each whole is being broken into 5 equal pieces. To indicate more than 1 whole, or $\frac{5}{5}$, the numerator must be bigger than the denominator 5. From problem 5(a), we know that $\frac{13}{5} = 2\frac{3}{5}$.

You try these *(Answers at end of chapter)*

Write each of the following improper fractions as mixed numbers.

(a) $\frac{17}{5}$ (b) $\frac{148}{7}$ (c) $\frac{52}{13}$

Problem 7 Write each of the following fractions or mixed numbers as decimal fractions.

(a) $\frac{4}{100}$ (b) $\frac{3}{5}$ (c) $\frac{6}{7}$ (d) $15\frac{2}{3}$

Solution (a) Because the denominator of this fraction is one of the decimal place-values, we can use the rules learned in Chapter 11 to write

Answer (a) $\frac{4}{100} = 0.04$

Solution (b) In this case, the denominator is not a decimal place-value, so we use the division meaning of fractions and a calculator to see that

$\frac{3}{5} = 3 \boxed{\div} 5 = 0.6$

(We represent the calculator operation keys to be punched by putting a box around them.)

Answer (b) $\frac{3}{5} = 0.6$

Solution (c) When we divide on the calculator, we get what is called a 'non-ending' decimal:

$\frac{6}{7} = 6 \boxed{\div} 7 = 0.8571428$

which would keep going on if our calculator had more places. In this text, we'll usually round off to the nearest hundredth, so

Answer (c) $\frac{6}{7} \doteq 0.86$

Solution (d) When we want to write a mixed number as a decimal fraction, the whole-number part remains as written followed by the decimal point and we change the fraction part to a decimal:

$\frac{2}{3} = 2 \boxed{\div} 3 = 0.6666666 \doteq 0.67$

Answer (d) $15\frac{2}{3} \doteq 15.67$

Note To change an improper fraction to a decimal fraction, we use the same method as for (a)–(c), but the answer you get on the calculator will have a whole number as well as decimal place values. For example,

$\frac{9}{7} = 9 \boxed{\div} 7 = 1.2857142 \doteq 1.29.$

You try these *(Answers at end of chapter)*

Write each of the following fractions or mixed numbers as decimal fractions.

(a) $\frac{6}{1,000}$ (b) $\frac{3}{4}$ (c) $\frac{8}{9}$ (d) $12\frac{1}{6}$

Applications

Problem 8 In the ninety-fifth Congress (1977) there were 292 Democratic Representatives and 143 Republican Representatives and no Representatives from other parties. About what fraction of the House of Representatives were Democrats?

Solution In general, fractions are not used in everyday life in as many situations as whole numbers, decimals or percentages. Mostly they are used to present data in a form which is more visual. This problem gives numerical data which can be described more clearly by an approximately reduced fraction. (When you learn percentages in Chapter 16, however, you may choose to represent such data in percentages.)

$$\frac{\text{Democrats}}{\text{all members of the House}} = \frac{292}{435}$$

This fraction cannot be exactly reduced (292 and 435 have

no common divisors other than 1). In real-life applications, however, we often round numbers and reduce approximately.

$$\frac{292}{435} \doteq \frac{300}{450}$$

Then reduce:

$$\frac{300 \div 50}{450 \div 50} = \frac{6 \div 3}{9 \div 3} = \frac{2}{3}.$$

So $\frac{292}{435} \doteq \frac{2}{3}$. (Surely $\frac{2}{3}$ gives a clearer visual picture than $\frac{292}{435}$ of the portion of Democrats.)

Answer For the ninety-fifth Congress, about $\frac{2}{3}$ of the Representatives were Democrats.

You try this *(Answer at end of chapter)*

About what fraction of the House were Republicans?

Problem 9 Assume you ran around a quarter-mile track 9 times. Write a mixed number and an improper fraction to represent the distance you ran.

Solution A quarter-mile track is a running path, usually the shape of a flattened circle, that measures $\frac{1}{4}$ of a mile.

Running around the track 9 times means you run 9 quarter-miles, or $\frac{9}{4}$ miles. Using the rule for changing improper fractions to mixed numbers, or using the logic that for every 4 laps (4 times around the track) you run 1 mile, we see that 9 laps around a quarter-mile track is $2\frac{1}{4}$ miles.

Answer $2\frac{1}{4}$ miles $= \frac{9}{4}$ miles.

You try this *(Answer at end of chapter)*

Write a mixed number and an improper fraction to represent the distance you would go if you ran around an eighth-of-a-mile track 19 times.

UNSOLVED PROBLEMS

(*Answers at end of chapter*)

Drill

1 Write two different fractions which represent the shaded portions of each of the diagrams.

(a) 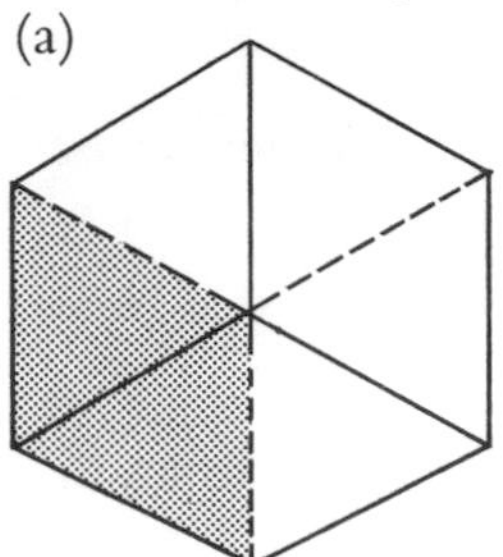

(b)

2 Write each of the following fractions in reduced form.

(a) $\frac{10}{25}$ (c) $\frac{15}{45}$

(b) $\frac{8}{20}$ (d) $\frac{35}{65}$

3 Write each of the following fractions in 'approximately reduced' form.

(a) $\frac{5}{11}$ (d) $\frac{22}{118}$

(b) $\frac{9}{28}$ (e) $\frac{46}{197}$

(c) $\frac{17}{25}$ (f) $\frac{150}{430}$

4 Given that each geometric figure represents 1 whole, write two expressions which represent the shaded portion of each of the following diagrams.

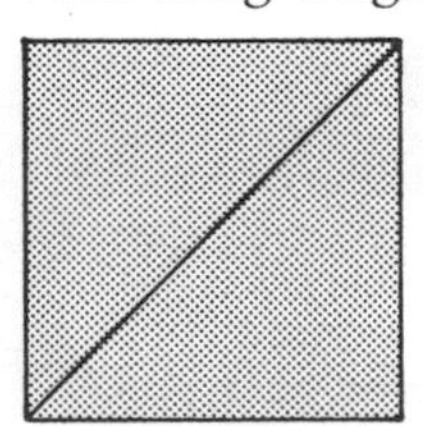 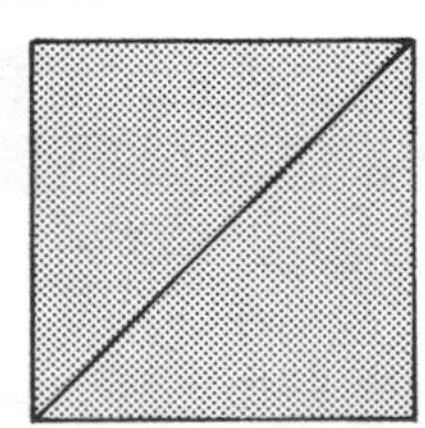 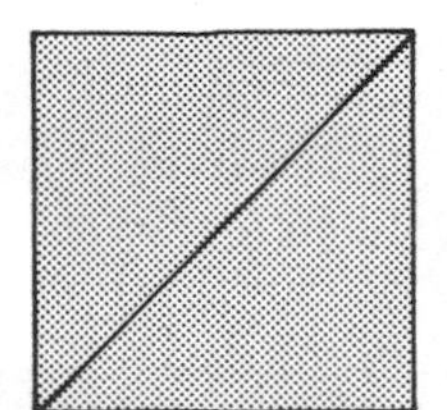 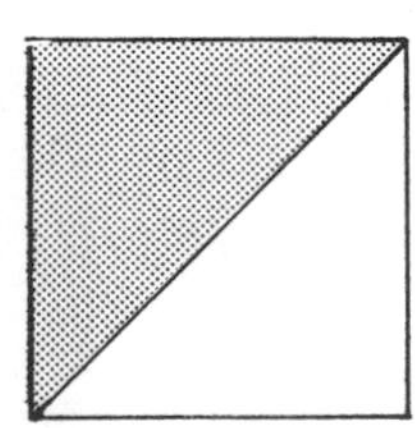

5 (Here, there are 4 different expressions)

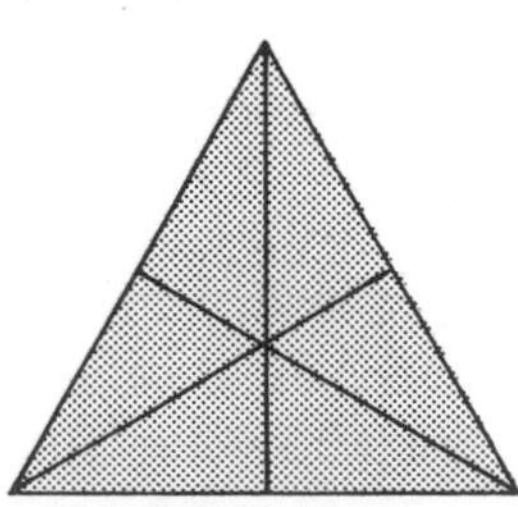 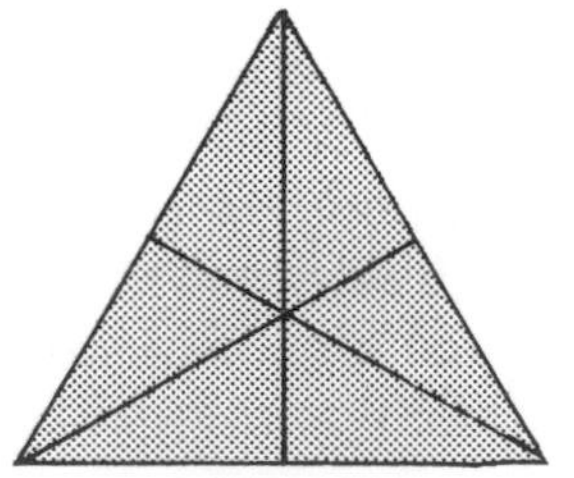 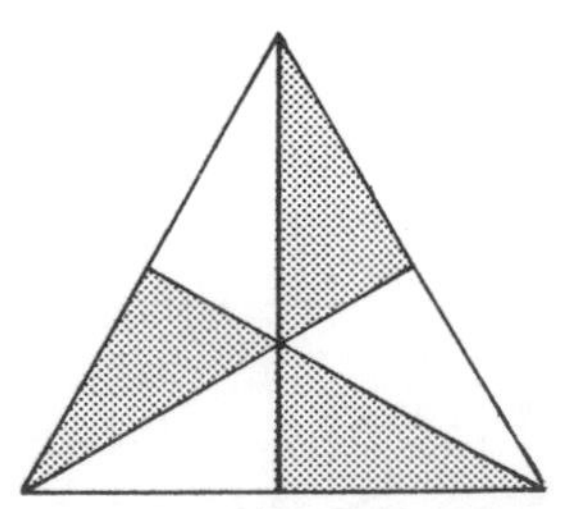

6 Draw a diagram which represents each of the following mixed-number improper-fraction equalities.

(a) $1\frac{4}{5} = \frac{9}{5}$ (b) $4\frac{3}{4} = \frac{19}{4}$

7 Write each of the following mixed numbers as an improper fraction.
(a) $3\frac{5}{8}$ (c) $12\frac{4}{5}$
(b) $10\frac{1}{6}$ (d) $12\frac{12}{12}$

8 Write each of the following improper fractions as mixed or whole numbers.
(a) $\frac{25}{8}$ (c) $\frac{25}{5}$
(b) $\frac{25}{4}$ (d) $\frac{144}{12}$

9 Write each of the following fractions or mixed numbers as decimal fractions.
(a) $\frac{32}{1,000}$ (d) $\frac{5}{7}$
(b) $\frac{5}{25}$ (e) $\frac{9}{14}$
(c) $\frac{12}{48}$ (f) $127\frac{5}{8}$

Thought

10 Find the error pattern in each of the following groups of fraction problems. Then do the problems correctly.

(a) Write each fraction in reduced form:

$\frac{16}{64} = \frac{1}{4}$ $\frac{19}{95} = \frac{1}{5}$ $\frac{62}{96} = \frac{2}{9}$

(b) Write each mixed number as an improper fraction:

$2\frac{2}{3} = \frac{6}{3}$ $3\frac{4}{5} = \frac{15}{5}$ $7\frac{3}{4} = \frac{28}{4}$

(c) Write each fraction as a mixed number:

$\frac{11}{5} = 2\frac{1}{11}$ $\frac{12}{7} = 1\frac{5}{12}$ $\frac{25}{8} = 3\frac{1}{25}$

(d) Write each fraction as a mixed number:

$\frac{5}{13} = 2\frac{3}{13}$ $\frac{6}{7} = 1\frac{1}{7}$ $\frac{7}{22} = 3\frac{1}{22}$

Applications

11 Ralph Nader's Health Research Group conducted a study of smoking policies in Washington, DC hospitals in 1976. They found

'Only a small fraction of those hospitals responding currently have regulations that adequately recognize and actively ensure the rights of all those who would prefer to breathe smoke-free air: 12 out of 21 hospitals do not solicit patient preferences regarding smoking in patient rooms prior to admission; 10 out of 21 hospitals sell cigarettes; 7 of 21 hospitals allow visitors to smoke in patient rooms, and 2 out of 21 even allow employees to smoke in patient rooms.'

(a) Find the fraction of hospitals involved in each type of abuse of non-smokers' rights; then reduce or approximately reduce that fraction to lowest terms.
(b) Survey some of the hospitals in your area and write a brief report about their smoking regulations; use approximate fractions to describe the data.

12 'Exploring gets but a fraction of extra oil profits, study says' (*Boston Globe*, 22 June 1981), discusses a consumer-group report which shows that, 'after years of demanding higher profits and less regulation because they said additional means were needed to seek out more oil, the nation's largest oil companies . . . are using their cash windfall to acquire competitors, buy more land than they can possibly explore efficiently and hoard vast amounts of cash'. Of course, as with all mergers, buying up existing oil firms increases the buyers' total oil reserves, but adds nothing to the country's supply of oil and gas, and does not subtract from unemployment by creating new jobs. 'From 1978 to 1980, America's 16 largest oil companies collected $29.3 billion in extra profits . . . but spent just $5.3 billion of it on domestic oil and gas exploration and production . . . at the same time, . . . [they] spent $11.1 billion to buy out smaller petroleum companies or acquire businesses unrelated to oil.'

(a) About what fraction of their extra profits were spent on domestic exploration and production?

(b) About what fraction of their extra profits were spent on acquiring companies?

13 In the 1980 Presidential election of Ronald Reagan, there were 160.5 million people of voting age, 110 million registered voters, and 83.7 million people who actually voted. About half of the people who voted cast ballots for Reagan.

(a) About what fraction of eligible voters are registered?

(b) About what fraction of eligible voters cast ballots for Reagan?

14 Use the following pie (or circle) graph, which visually shows the 'fractions', to approximate the numerical fraction of the US Federal Research and Development spending in 1976 for each of the given categories.

US Federal R&D spending (1976)

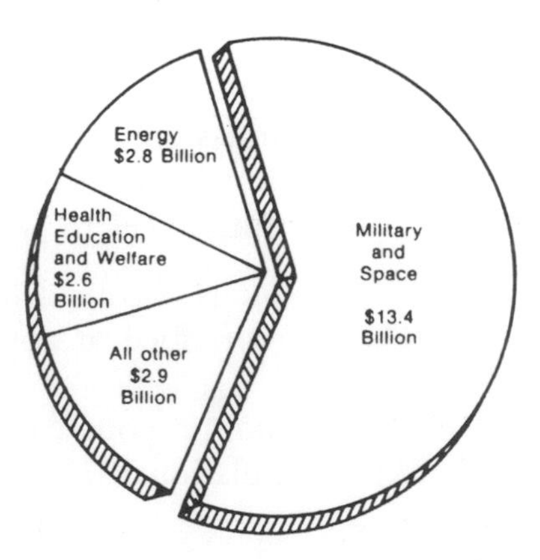

Source: National Science Foundation/Center for Defense Information

15 'The economics of racism in South Africa' (*New York Times*, 19 September 1976) uses numbers to illustrate

> 'the extent to which the country's 4.2 million whites have been sustained by the labor of its 18 million blacks . . . the Government's own statistics . . . suggest that scarcely any sector of the economy could function for long without black workers. Of the 4.9 million full time workers registered last year . . . 2.7 million were black, 1.5 million were white, 527,740 were of mixed race, and 181,066 were Asian . . . In gold mining, the pillar of the economy, the labor force of 380,091 included 341,575 blacks. Figures for other sectors show a similar ratio: coal mining, 190,596 (134,012); construction, 446,086 (309,094); textiles, 97,628 (66,939); and food processing, 159,577 (103,286) . . . The Government's manpower survey lists the number of blacks employed in the various professions . . . 11 engineers, 31 chemists, 63 lawyers, 83 university teachers, 85 doctors, 528 journalists and writers . . . 74,980 [school teachers] . . . '

(a) About what fraction of the workers in each of the industries listed are black?
(b) About what fraction of all black workers are professionals?
(c) In contrast to their importance in the economy, 'Average annual income among urban whites is estimated to be about $6,000; among urban blacks . . . between $860 and $1,600 . . . the Johannesburg Chamber of Commerce [estimates] that the subsistence income for a family of five in Soweto in May 1975 was $136 a month . . . 5.6 million blacks [live] in urban areas . . . In 1973, the average annual income among black farm laborers was $250.' Create and solve two maths problems, using this information, whose solution involves approximately reducing fractions.

16 *Portrait of Inequality: Black and White Children in America*, by Marion Wright Edelman (Children's Defense Fund, 1980), documents that, in spite of some success, 'Black children, youths, and families remain worse off than whites in every area of American life . . . ' In education, for example,

> 'A Black child is almost twice as likely as a White child to grow up in a family whose head did not complete high school . . . a White child is four times as likely as a Black child to grow up in a family headed by a college graduate . . . Black children are suspended from school at twice the rate for White children. Over half a million of the 1.8 million students suspended were Black . . . Out of some 1,264,000 American children aged 7 to 17 who were not enrolled in any school in 1976, 172,000 were Black.'

(a) About what fraction of suspended students were black?

(b) About what fraction of students aged 7 to 17 not enrolled in school were black?

(c) Often, to compare data from groups of different sizes we calculate *rates*, which are fractions with denominators of 100, 1000, 100,000, etc. For example, it may not be immediately evident that the data in the table below show a much worse situation for black youth.

Number of school drop-outs and high-school graduates aged sixteen to twenty-four, 1977–8

	White	Black
Drop-outs	640,000	172,000
High-school graduates	2,747,000	347,000

Source: US Department of Labor, Bureau of Labor Statistics Press

In order to clarify these data, the Children's Defense Fund calculated the rate of drop-outs per hundred graduates for whites and blacks. Find the respective rates and discuss why they present a clearer picture of the data.

17 Write a mixed number or a whole number, and an improper fraction, for each of the following measurements:

(a) Running 15 laps on a quarter-mile track.

(b) Running 20 laps on a half-mile track.

(c) Running 3 miles on an eighth-of-a-mile track.

(d)

(e)

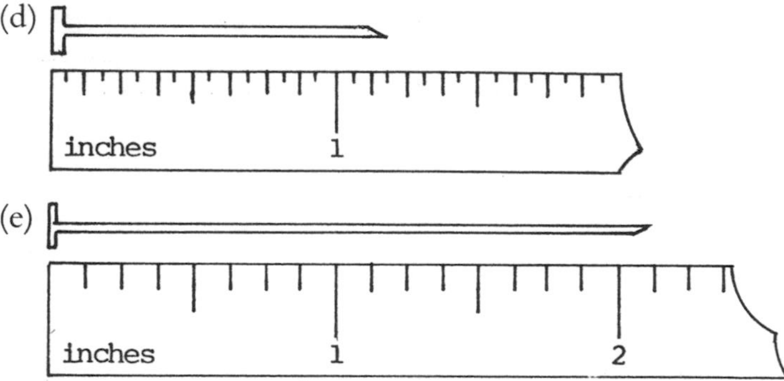

(f) The number of hours elapsed from 7.30 a.m. to 3.30 p.m.

18 The following bar graph from *Progressive Agenda* (August 1986) updates the information in Chapter 13 (solved problem 8), and presents it in a different form.

(a) What conclusions can you draw from the information in the graph?

(b) How were the decimal fractions for each bar computed?

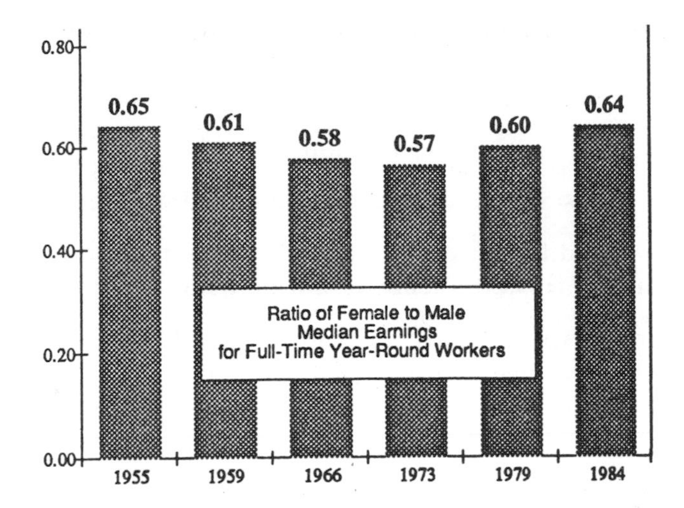

19 A 1974 Committee for Nuclear Responsibility (CNR) flyer reported that by the year 2000 the Atomic Energy Commission had expected to license for operation 1,000 nuclear power plants. (Three-Mile Island, skyrocketing construction costs, and popular protest have slowed them down.) CNR uses a numerical argument to show the deception of 'low' accident-probability figures: '*If* we permit a thousand plants to operate, and *if* the probability of a major accident were really as low as one in a million per reactor per year, then the probability of a major accident during the 40-year lifespan of all the plants would be about one chance in 25!' Show how the $\frac{1}{25}$ probability was computed.

20 Create and solve a maths problem, based on some current information you find in a newspaper, in a flyer from some community organizers, or the like, whose solution involves reducing or approximately reducing fractions.

Recreation

21 The decimal fractions for $\frac{1}{7}, \frac{2}{7}, \frac{3}{7}, \frac{4}{7}, \frac{5}{7}, \frac{6}{7}$ form a fascinating pattern. Use a calculator and long division to write these fractions as decimals with more than 12 places, and figure out how the following diagram describes the cyclic pattern of the sevenths:

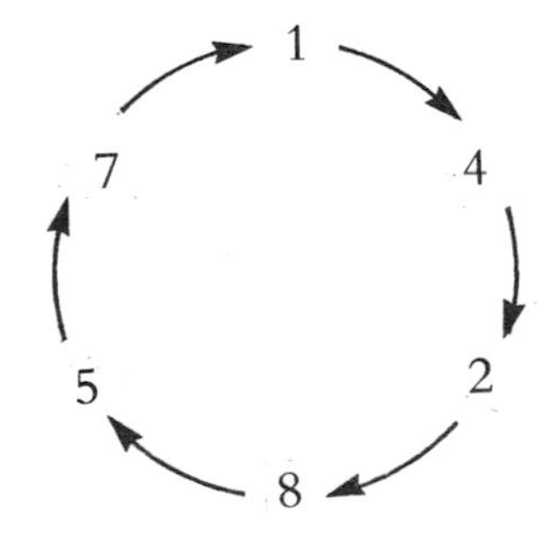

REVIEW QUIZ (different ways of writing fractions and mixed numbers) *(Answers at end of chapter)*

1 Write in 'approximately' reduced form $\frac{28}{73}$.

2 Write $3\frac{6}{7}$ as an improper fraction.

3 Write $13\frac{7}{9}$ as a decimal fraction, rounded to the nearest hundredth.

4 Write two different expressions to represent the distance covered in running 15 laps on a tenth-of-a-mile track.

5 According to the study cited in unsolved problem 12, the sixteen largest oil companies spent $2.8 billion of their $29.3 billion in extra profits on foreign exploration and production. About what fraction of their extra profits did they spend on foreign exploration and production?

ANSWERS: CHAPTER 14

Solved problems ('you try these')

1 $\frac{2}{3} = \frac{4}{6}$

2 (a) $\frac{2}{5}$ (b) $\frac{8}{45}$ (c) $\frac{2}{5}$

3 $\frac{316}{897} \doteq \frac{300 \div 300}{900 \div 300} = \frac{1}{3}$

4 $1\frac{2}{3} = \frac{5}{3}$

5 (a) $\frac{17}{6}$ (b) $\frac{112}{12}$ (c) $\frac{6}{1}$

6 (a) $3\frac{2}{5}$ (b) $21\frac{1}{7}$ (c) 4

7 (a) 0.006 (b) 0.75 (c) 0.89 (d) 12.17

8 $\frac{143}{435} \doteq \frac{150}{450} = \frac{1}{3}$ or since $\frac{2}{3}$ are Democrats, the remaining $\frac{1}{3}$ are Republicans.

9 $\frac{19}{8} = 2\frac{3}{8}$ miles.

Unsolved problems

1 (a) $\frac{1}{3} = \frac{2}{6}$ (b) $\frac{1}{2} = \frac{2}{4} = \frac{4}{8}$

2 (a) $\frac{2}{5}$ (b) $\frac{2}{5}$ (c) $\frac{1}{3}$ (d) $\frac{5}{8}$

3 (a) $\frac{5}{11} \doteq \frac{5}{10} = \frac{1}{2}$
(b) $\frac{9}{28} \doteq \frac{10}{30} = \frac{1}{3}$
(c) $\frac{17}{75} \doteq \frac{15}{75} = \frac{1}{5}$
(d) $\frac{22}{118} \doteq \frac{20}{120} = \frac{1}{6}$
(e) $\frac{46}{197} \doteq \frac{50}{200} = \frac{1}{4}$
(f) $\frac{150}{430} \doteq \frac{150}{450} = \frac{1}{3}$

4 $3\frac{1}{2} = \frac{7}{2}$

5 $2\frac{3}{6} = \frac{14}{6}$ or $2\frac{1}{2} = \frac{2}{5}$

7 (a) $\frac{29}{8}$ (b) $\frac{61}{6}$ (c) $\frac{64}{5}$ (d) $\frac{156}{12}$ (note: $12\frac{12}{12} = 12 + 1 = 13$)

8 (a) $\frac{25}{8} = 3\frac{1}{8}$ (b) $\frac{25}{4} = 6\frac{1}{4}$ (c) $\frac{25}{5} = 5$ (d) $\frac{144}{12} = 12$

9 (a) 0.032
(b) 0.2
(c) 0.25
(d) $\frac{5}{7} \doteq 0.71$
(e) $\frac{9}{14} \doteq 0.64$
(f) 127.625

10(a) Interestingly, although this method is wrong, the answers (coincidentally) are correct! But the method rarely gives the correct answer, because crossing out the same *digit* from numerator and denominator is only meaningful when that digit is zero in the one's place, where crossing out a zero is the same as dividing by ten; crossing out other digits has no consistent meaning. For example, crossing out the 6 in 16 to leave 1 does not correspond to any mathematical operation; $\frac{16}{64} = \frac{1}{4}$ because $\frac{16 \div 16}{64 \div 16} \frac{1}{4}$ not because you cross out 6's.

(b) This person forgot to add the numerator. $2 = \frac{6}{3}$, $2\frac{1}{4} = \frac{8}{3}$

(c) This person was confused about the denominator, for $\frac{11}{5} = 2\frac{1}{5}$; s/he used the numerator, instead of the denominator, of the improper fraction for the denominator of the mixed number.

(d) This person divided the numerator into the denominator – these fractions are not improper. $\frac{5}{13} < \frac{13}{13} = 1$, so $\frac{5}{13}$ surely doesn't equal $2\frac{2}{13}$. You can only divide denominator into numerator. These fractions are not greater than 1 and can't be written, therefore, as mixed numbers.

11(a) $\frac{12}{21} = \frac{4}{7}$; $\frac{7}{21} = \frac{1}{3}$; $\frac{2}{21}$ is in lowest terms, but $\frac{2}{21} \doteq \frac{1}{10}$.
The report did not reduce these fractions because there were only 21 hospitals involved in the survey and it is less misleading to give the exact numbers when the total surveyed is small. In addition to the adverse health effects of smoking on non-smoking patients, the survey also revealed evidence of hospital fires which had arisen in Washington DC as a result of smoking.

12 (a) $\dfrac{5.3\ \cancel{\text{billion}}}{29.3\ \cancel{\text{billion}}} = \dfrac{5.3}{29.3} \doteq \dfrac{5}{30} = \dfrac{1}{6}$

(b) $\dfrac{11.1\ \cancel{\text{billion}}}{29.3\ \cancel{\text{billion}}} \doteq \dfrac{10}{30} = \dfrac{1}{3}$

13 (a) $\dfrac{110\ \cancel{\text{million}}}{160.5\ \cancel{\text{million}}} \doteq \dfrac{11\cancel{0}}{16\cancel{0}} = \dfrac{11}{16} \doteq \dfrac{10}{16} = \dfrac{5}{8}$

(b) half of 83.7 $\doteq$ half of 84 = 42 and $\dfrac{42\ \cancel{\text{million}}}{160.5\ \cancel{\text{million}}} \doteq \dfrac{40}{160} = \dfrac{1}{4}$

14 The total is (\$13.4 + \$2.8 + \$2.6 + \$2.9) billion = \$21.7 billion

$\dfrac{\text{military}}{\text{total}} = \dfrac{13.4\ \cancel{\text{billion}}}{21.7\ \cancel{\text{billion}}} \doteq \dfrac{14}{21} = \dfrac{2}{3}$;

$\dfrac{\text{energy}}{\text{total}} = \dfrac{\text{health, etc.}}{\text{total}} \doteq \dfrac{\text{all other}}{\text{total}} \doteq \dfrac{3}{21} = \dfrac{1}{7}$

15(a) gold mining: $\dfrac{\text{blacks}}{\text{total}} = \dfrac{341{,}575}{380{,}081} \doteq \dfrac{35\cancel{0},\cancel{000}}{40\cancel{0},\cancel{000}} = \dfrac{35}{40} = \dfrac{7}{8}$
Note that you might have rounded in other ways, for example:

$\dfrac{34\cancel{0},\cancel{000}}{38\cancel{0},\cancel{000}} = \dfrac{34}{38} = \dfrac{17}{19} \doteq \dfrac{15}{20} = \dfrac{3}{4}$.

Since $\frac{3}{4} = \frac{6}{8}$, you get different, but reasonably close fractions. Remember, these are only approximations to give a rough picture of the proportion of black workers in each industry.

Coal mining: $\dfrac{134{,}012}{190{,}596} \doteq \dfrac{15\cancel{0},\cancel{000}}{20\cancel{0},\cancel{000}} = \dfrac{15}{20} = \dfrac{3}{4}$;

Construction: $\dfrac{309{,}094}{446{,}086} \doteq \dfrac{30\cancel{0},\cancel{000}}{45\cancel{0},\cancel{000}} = \dfrac{30}{45} = \dfrac{2}{3}$;

Textiles: $\dfrac{66{,}939}{97{,}628} \doteq \dfrac{7\cancel{0},\cancel{000}}{10\cancel{0},\cancel{000}} = \dfrac{7}{10}$;

Food processing: $\dfrac{103{,}286}{159{,}577} \doteq \dfrac{10\cancel{0},\cancel{000}}{16\cancel{0},\cancel{000}} = \dfrac{10}{16} = \dfrac{5}{8}$

(b) Assuming that all the blacks in professions were listed,

$\dfrac{\text{black professionals}}{\text{total number black workers}} = \dfrac{11 + 31 + 63 + 83 + 85 + 528 + 74{,}980}{2.7 \text{ million}} =$

$\dfrac{75{,}781}{2.7 \text{ million}} \doteq \dfrac{75,\cancel{000}}{2{,}700,\cancel{000}} = \dfrac{75}{2{,}700} = \dfrac{1}{36}$ (since $2{,}700 \div 75 = 36$).

16(a) $\dfrac{5\cancel{00},\cancel{000}}{1{,}8\cancel{00},\cancel{000}} = \dfrac{5}{18} \doteq \dfrac{5}{20} = \dfrac{1}{4}$

(b) $\frac{172{,}000}{1{,}264{,}000} \doteq \frac{200{,}000}{1{,}200{,}000} = \frac{2}{12} = \frac{1}{6}$

Note that the rounding is not according to the strict rules. Even though it is closer to 1,300,000 I rounded 1,264,000 to 1,200,000 because I saw that I'd get $\frac{2}{12}$ which reduces to $\frac{1}{6}$. If I had rounded to 1,300,000 I would just round again later:

$\frac{200{,}000}{1{,}300{,}000} = \frac{2}{13} \doteq \frac{2}{12} = \frac{1}{6}$

(c) White: $\frac{640{,}000}{2{,}747{,}000} = \frac{?}{100}$ is what is asked for. Since $2{,}747{,}000 \div 100 = 27{,}470$ we divide the numerator by the same number to reduce to an equivalent fraction with denominator equal to 100:

$\frac{640{,}000 \div 27{,}470}{2{,}747{,}000 \div 27{,}470} \doteq \frac{23.3}{100}$. black: $\frac{172{,}000 \div 3{,}470}{347{,}000 \div 3{,}470} \doteq \frac{49.6}{100}$

The figures for drop-outs per 100 graduates show that even though there are about four times as many white drop-outs as black drop-outs, the proportion of black drop-outs compared to black graduates is much higher than the corresponding rate for whites. Since $\frac{49.6}{100} \doteq \frac{50}{100} = \frac{1}{2}$, there is about 1 black drop-out for every 2 black high-school graduates. (The Children's Defense Fund book suggests many concrete political tasks for black communities in the fight to change this situation.)

17(a) $\frac{15}{4} = 3\frac{3}{4}$ miles
(b) $\frac{20}{2} = 10$ miles
(c) $3 = \frac{24}{8}$ miles = 24 laps
(d) $1\frac{3}{16} = \frac{19}{16}$ inches
(e) $2\frac{1}{8} = \frac{17}{8}$ inches
(f) $8 = \frac{16}{2}$ (or $\frac{32}{4}$) hours

18(a) As the article states, 'Women workers earn less than two-thirds of their male counterparts [$\frac{2}{3} \doteq 0.67$]; and, this ratio has fluctuated within a relatively narrow band for more than 30 years.'

(b) The ratio, or fraction $\frac{\text{male earnings}}{\text{female earnings}}$ was formed and then rewritten as a decimal by dividing: female earnings ÷ male earnings.

19 1 in a million per reactor per year, if there are 1,000 reactors usable for 40 years each, means $1{,}000 \times 40 = 40{,}000$ reactor-years.

$\frac{1}{1{,}000{,}000}$ for each reactor year means $\frac{40{,}000}{1{,}000{,}000}$ for the 40,000 reactor-years.

$\frac{40{,}000}{1{,}000{,}000} = \frac{4}{100} = \frac{1}{25}$. The flyer adds to this strictly numerical argument by noting that figures on probabilities of nuclear accidents ignore: (a) that error or malice could instantly reduce the catastrophe-odds to near certainty; (b) that 'experts' have very little experience operating nuclear hardware; and (c) that numerous 'surprises' have shown that nuclear engineers are failing to foresee all the design problems in nuclear safety-systems (and this was written before Three-Mile Island and Chernobyl!). When computing these probabilities, nuclear 'experts' assume no error, no malice, experienced operators, and properly designed safety-systems!

Review quiz

1 $\frac{28}{73} \doteq \frac{30}{75} = \frac{2}{5}$

2 $\frac{27}{7}$

3 13.78

4 $\frac{15}{10} = 1\frac{1}{2}$ miles

5 $\frac{\$2.8 \text{ ~~billion~~}}{\$29.3 \text{ ~~billion~~}} \doteq \frac{3}{30} = \frac{1}{10}$

15 Comparing and rounding fractions

According to *Our Bodies, Ourselves* (Boston Women's Health Book Collective),

> each of the currently available [birth control] methods has disadvantages as well as advantages. Some methods we find to be a nuisance, others may make us sick. Many may have long-term dangers still unknown. The choice usually involves deciding where we are willing to make a compromise. We can weigh whether effectiveness, safety, or convenience matters more to us, and . . . what method we feel the most comfortable with and which one we will use the most consistently.

The following table on contraceptive effectiveness gives 'actual use failure rate', based on records of actual use of the method over time. 'Actual failure rates include accidents such as forgetting a pill . . . removing the diaphragm within six hours of intercourse . . . Drug company literature tends to present only the theoretical rate [based on perfect use of the method].'

Approximate failure rate of various contraceptives

Method	Actual use failure rate
Hysterectomy	1 out of 1,000,000
Vasectomy	15 out of 10,000
Oral contraceptives (combined)	35 out of 1,000
Condom and spermicide	5 out of 100
IUD (intra-uterine device)	6 out of 100
Condom	175 out of 1,000
Diaphragm	225 out of 1,000
Spermicidal foam	30 out of 100
Rhythm (calendar)	35 out of 100
Chance (sexually active with male partner)	80 out of 100

Note: A '5 out of 100' failure rate means that 'studies in the past have shown that 5 women out of 100 using that particular method have become pregnant in one year'.

If we wanted to compare the effectiveness of, say, the combined oral contraceptives ($\frac{35}{1,000}$) to the IUD ($\frac{6}{100}$), we would have to be able to tell which fraction was greater. We can use a calculator and the division meaning of fractions, to change these fractions to decimals: $\frac{35}{1,000} = 0.035$ and $\frac{6}{100} = 0.06$ and compare the decimals: $0.06 > 0.035$, so the failure rate with the IUD is greater than the failure rate with the pill.

SOLVED PROBLEMS

(Answers at end of chapter)

Drill

Problem 1 Replace the question mark with the correct comparison symbol:

$$\frac{6}{7} \ ? \ \frac{9}{11}$$

Solution Using the division meaning of fractions, we write each fraction as a decimal,

$\frac{6}{7} = 6 \boxed{\div} 7 \doteq 0.86$

$\frac{9}{11} = 9 \boxed{\div} 11 \doteq 0.82$

and then use the rules learned in Chapter 12 for comparing decimals:

$0.86 > 0.82$, so

Answer $\frac{6}{7} > \frac{9}{11}$.

You try this *(Answer at end of chapter)*

Replace the question mark with the correct comparison symbol.

$\frac{2}{3} \ ? \ \frac{5}{8}$

Problem 2 Round $5\frac{6}{13}$ to the nearest whole number.

Solution Again, first change $5\frac{6}{13}$ to a decimal:

$\frac{6}{13} = 6 \boxed{\div} 13 \doteq 0.46$

So, $5\frac{6}{13} \doteq 5.46$ which, using the rules learned in Chapter 12, rounds to 5.

Answer $5\frac{6}{13}$ rounded to the nearest whole number is 5.

You try this *(Answer at end of chapter)*

Round $19\frac{9}{15}$ to the nearest whole number.

UNSOLVED PROBLEMS

(*Answers at end of chapter*)

Drill

1 Replace the question mark with the correct comparison symbol.

(a) $\frac{1}{3}$? $\frac{2}{3}$
(b) 5 ? $\frac{6}{7}$
(c) $2\frac{1}{2}$? $7\frac{1}{2}$
(d) $\frac{14}{3}$? $3\frac{3}{4}$
(e) $15\frac{7}{12}$? $15\frac{3}{4}$
(f) $\frac{6}{13}$? $\frac{7}{15}$

2 Round each of the following mixed numbers to the nearest whole number.

(a) $2\frac{1}{3}$
(b) $15\frac{3}{5}$
(c) $10\frac{9}{10}$
(d) $6\frac{3}{4}$

Thought

3 (a) Put the following fractions in size place and form a generalization about comparing the size of fractions with numerators equal to 1.

$\frac{1}{2}$; $\frac{1}{7}$; $\frac{1}{3}$; $\frac{1}{5}$; $\frac{1}{6}$; $\frac{1}{4}$; $\frac{1}{8}$

(b) Put the following fractions in size place and form a generalization about comparing the size of fractions with the same denominator.

$\frac{2}{100}$; $\frac{62}{100}$; $\frac{35}{100}$; $\frac{12}{100}$; $\frac{1}{100}$; $\frac{98}{100}$

4 Read '0 = 1 = $\frac{8}{12}$'. (*The Mathematics Teacher*, January 1983)

(a) What is the main point?
(b) Explain why $\frac{8}{12}$ is rounded to 1 when rounding 'to the nearest integral value'.
(c) Explain why $\frac{8}{12}$ is rounded to 0 when rounding 'to the nearest multiple of 2'.

'0 = 1 = $\frac{8}{12}$?

'One side of the label of a Pepsi Light eight-ounce can reports that it contains 1 calorie. Another portion of the label reports that it contains no calories. The truth is that it contains $\frac{8}{12}$ of a calorie.

'One label rounds the $\frac{8}{12}$ to the nearest integral value of 1. For the other label, the Food and Drug Administration regulations require that products must be labeled with caloric content per serving and that the calories per serving must be rounded to the nearest multiple of 2. Since an eight-ounce can is one serving, the $\frac{8}{12}$ is rounded to 0, the nearest multiple of 2.'

5 Use the following letter to the editor of the *New York Times* (1 March 1977) to answer the following questions:

(a) What is the main point of this letter?
(b) Give a specific example from your own food shopping experiences where 'fractured fractions' have been a problem.

'Groceries: beholding fractured fractions

'To the Editor:

'There is a mathematical madness among manufacturers of packaged foods. Studying a recent neighborhood grocery circular, I found the following weights and measurements advertised in as many packages: 13 oz, 16 oz, 15 oz, $11\frac{1}{2}$ oz, 1 lb 2 oz, 6.4 oz, 9 oz, 13 oz, $7\frac{1}{4}$ oz, $9\frac{1}{2}$ oz, 7 oz, 1 lb 13 oz, 3 lb 1 oz, $14\frac{1}{2}$ oz, $3\frac{1}{2}$ oz, 29 oz, 1 pt 4 oz, 1 qt 14 oz, *ad nauseam*.

'You can bet your $3\frac{1}{2}$ oz brain and $11\frac{1}{4}$ oz heart that this case of fractured fractions has no connection to any efforts by the manufacturers to clarify costs to the consumer. What a consummate waste!

'Charles H. Lavigne
'Bayside, NY, Feb. 20, 1977'

6 Use the table at the beginning of this chapter to answer the following questions:
(a) Compare the effectiveness of the condom to the effectiveness of the diaphragm.
(b) Compare the effectiveness of a vasectomy to the effectiveness of the pill.
(c) What is the least effective method of birth control?

7 Another important consideration in choosing a birth-control method is safety. Use this table adapted from *Our Bodies, Ourselves* to answer the following questions.

(Approximate) Risk of death for women in the age group fifteen to forty-five (USA, 1973)

Legal therapeutic abortion	32 out of 1,000,000 women
Oral contraceptives	135 out of 10,000,000
IUD	6 out of 10,000

(a) Which of the listed contraceptive methods has the greatest risk of death?
(b) Which of the listed contraceptive methods has the smallest risk of death?

8 Use the following letter to the editor *Consumer Reports*, June 1978, to answer the following questions:
(a) What is the main point of this letter?
(b) If canned vegetables cost 3 cans for $1, *exactly* how much does each can cost?
(c) Why does the computer round the exact cost to 34 cents? (Is this consistent with the rules for rounding fractions?)
(d) In what other real-life instances are the rules for rounding not followed?

'Adding fractions

'Our local Giant supermarket recently installed computerized checkout counters, and I have noticed an interesting fact of these devices. Suppose I buy, for example, a can of peas, a can of corn, and a can of green beans, each marked 3/$1. The wonderful printed receipt will show the name of each product and a charge of 34 cents for each. I pay $1.02. Several cashiers said that on old registers they would ring up $1 for the three items. The computer programming necessary to mimic the action of the cashiers could easily be accomplished. The existing programs already give the computer the ability to recognize three of the same items (green beans, for example), at 3/$1 and ring them up at 34 cents, and 33 cents. It seems a shame that a device which is supposed to reduce grocery costs ends up costing customers money in the short run.'

'Baltimore 'D.A.P.'

9 Since the introduction of the instant lottery game in Massachusetts, sales of weekly lottery tickets have decreased significantly, supporting the contention that gamblers prefer a game with instant feedback.

Massachusetts lottery sales

Year	Total	Weekly	Instant	Numbers
1974	$693M*	$676M	$17M	–
1975	$936M	$649M	$238M	$50M
1976	$1.2B*	$449M	$642M	$154M
1977	$1.7B	$326M	$755M	$618M
1978	$2.0B	$354M	$726M	$921M
1979	$2.0B	$179M	$509M	$1.4B

*M = millions; B = billions.

Source: Massachusetts Lottery Commission (from a *Boston Globe* article, 1 May 1980)

(a) In which year was the greatest fraction of lottery money spent on the weekly lottery?

(b) In which year was the greatest fraction of lottery money spent on the instant lottery?

(c) Why don't the weekly and instant and number columns add up exactly to the total column?

(d) Create and solve a maths problem, based on the information in the given chart, whose solution involves comparing fractions.

10 What can you conclude from each of the following facts (taken from 'How poverty breeds overpopulation' (Commoner, 1975):

(a) '. . . at around 1800, Sweden had a high birth rate (about $33/1,000$ [33 births per 1,000 people in Sweden]), but since the death rate was equally high, the population was in balance. Then as agriculture, and, later, industrial production advanced, the death rate dropped until, by the mid-nineteenth century, it stood at about $\frac{20}{1,000}$ . . . Then, however, the birth rate began to drop, gradually narrowing the gap until in the mid-twentieth century it reached about $\frac{14}{1,000}$, when the death rate was about $\frac{10}{1,000}$.'

(b) '. . . in a number of Latin American and Asian countries . . . as infant mortality drops from $\frac{80}{1,000}$ to about $\frac{25}{1,000}$ (the figure characteristic of most developed countries), the birth rate drops sharply from $\frac{45}{1,000}$ to about 15 to 18 per 1,000.'

(c) We can use gross national product (GNP) per person (the total money value of goods and services produced in a year in a given country divided by the population of the country) as a comparative measure of standard of living (although it's a problematic measure, since it does not take into account quality of life or unequal distribution of wealth among the population). The poorest countries (GNP per capita $< \$500$ per year, in 1969 US dollars) have the highest birth rates, about $\frac{45}{1,000}$. When GNP per capita per year exceeds \$500 the birth rate drops sharply, reaching about $\frac{20}{1,000}$ at \$750–\$1,000. Most of the nations in North America, Oceania, Europe, and the USSR have about the same low birth rates – 15 to 18 per 1,000 – but their GNPs per person per year range all the way from Greece (\$941; birth rate $\frac{17}{1,000}$) through Japan (\$1,626; birth rate $\frac{18}{1,000}$) to the richest country of all, the USA (\$4,538; $\frac{18}{1,000}$).

REVIEW QUIZ (comparing and rounding fractions)

(Answers at end of chapter)

1 Replace the question mark with the correct comparison symbol.
(a) $\frac{2}{3}$? $\frac{5}{7}$ (b) $4\frac{5}{8}$? $4\frac{31}{56}$

2 Round to the nearest whole number: $99\frac{9}{17}$.

3 Use the table in problem 9 to answer these questions:
(a) Was the fraction of lottery money spent on the weekly game greater in 1975 or 1976?
(b) Was the fraction of lottery money spent on numbers greater in 1975 or 1976?

ANSWERS: CHAPTER 15

Solved problems ('you try these')

1 $\frac{2}{3} > \frac{5}{8}$

2 20

Unsolved problems

1(a) $\frac{1}{3} < \frac{2}{3}$
 (b) $5 > \frac{6}{7}$
 (c) $2\frac{1}{2} < 7\frac{1}{2}$
 (d) $\frac{3}{14} > 3\frac{3}{4}$
 (e) $15\frac{7}{12} < 15\frac{3}{4}$
 (f) $\frac{6}{13} < \frac{7}{15}$

2(a) 2
 (b) 16
 (c) 11
 (d) 7

3(a) $\frac{1}{8} < \frac{1}{7} < \frac{1}{6} < \frac{1}{5} < \frac{1}{4} < \frac{1}{3} < \frac{1}{2}$. If the numerators of two or more fractions are equal to 1, then the fraction with the largest denominator is the smallest fraction (this makes sense since the larger denominator means the pie is being cut into more pieces, so each piece will be smaller).
 (b) $\frac{2}{100} < \frac{2}{100} < \frac{12}{100} < \frac{35}{100} < \frac{62}{100} < \frac{98}{100}$. If the denominators of a group of fractions are the same, then the smaller fraction is the one with the smaller numerator.

4(a) Pepsi doesn't really have *no* calories; the advertisers say that because technically they can round $\frac{8}{12}$ to 0.
 (b) $\frac{8}{12} = 8 \div 12 \doteq 0.67 \doteq 1$ (to the nearest integer, that is, whole number).
 (c) 0.67 is closer to 0 than to 2, so under those rounding rules 0.67 rounds to 0.

5(a) Manufacturers make it very difficult for people to compare values by packaging goods in fractional amounts.

6(a) The condom is more effective than the diaphragm (it results in pregnancy less often than the diaphragm).
 (b) $\frac{35}{1,000} = 0.035$ and $\frac{15}{1,000} = 0.0015$, so the vasectomy is more effective.
 (c) Chance (this certainly fits with our intuition!).

7 Legal therapeutic abortion: 0.000032, oral contraceptives: 0.0000135, IUD: 0.0006, so the death rate from IUD is greatest and from legal therapeutic abortion is smallest. *Our Bodies, Ourselves* adds that 'it can be argued that taking the pill is safer than risking pregnancy. The death rate due to blood clots in women on the pill is estimated at 3 per 100,000; the death rate during pregnancy and delivery is 14 to 17 per 100,000. But . . . the figures are misleading. The risk of death in childbirth is not the same for every woman, and varies with social class, race, and age . . . For some women the risk of death from the pill may be as great as the risk of death from childbirth . . . A further consideration here is that death rates are no indication of the significant number of women who have been hospitalized with crippling strokes or other non-fatal blood clots, who develop diabetes on the pill, or who become debilitatingly depressed.'

8(a) Stores always 'round up', presumably to make more profit. But a cashier only does this when you, say, buy one item and the price is '3 for $1'. The computer evidently can't tell whether you've bought just one item or three items and it bills for each one separately as if you had only bought that one. Because $\frac{1}{3} < \frac{1}{2}$, the rules for rounding say $33\frac{1}{2}$ rounds to the nearest whole to 33.
 (b) $\frac{100}{3} = 33\frac{1}{3}$ per can.

9(a) $\frac{\text{weekly}}{\text{total}}$, 1974: $\frac{676\text{M}}{693\text{M}} \doteq 0.975$; 1975: 0.693; 1976: $\frac{449\text{M}}{1.2\text{B}} = \frac{449\text{M}}{1,200\text{M}}$ $\doteq 0.374$; 1977: 0.192; 1978: 0.177 (interestingly this ratio is *exact*: $\frac{354}{2,000} = 0.177$); 1979:

0.0895 (this figure is also exact). The fraction decreases each year, the greatest was 1974.

(b) $\frac{\text{instant}}{\text{total}}$, 1974: 0.025; 1975: 0.254; 1976: 0.535 (also an exact ratio); 1977: 0.444; 1978: 0.363 (also exact); 1979: 0.2545 (exact). So these fractions have varied considerably, reaching a high in 1976.

(c) The totals are rounded. For example in 1978, weekly + instant + numbers = 2,001M = 2.001B, rounded in the table to 2B. The article reports that 'Lotteries are especially popular in Third World nations. "In many parts of Africa, South America, and Asia" said Richard Schneiker of Scientific Games [which developed the instant lottery tickets], "the standard of living is so low that the only chance people have to get out of poverty, their only shot at a decent life, is by winning big in a lottery."' Fortunately, people in many Third World countries don't follow this advice; they feel socialism gives them better odds for a decent life! (The odds in a lottery, by the way, are frequently as high as 35 million to 1.)

10(a) 'Under the influence of a constantly rising standard of living the population moved, with time, from a position of balance at a high *death rate* to a new position of near-balance *at a low death rate*. But in between the population increased considerably.'

(b) This is evidence that a similar generalization holds in developing countries. As the standard of living rises and living conditions, such as nutrition, improve, the death rate drops. After time, the birth rate begins voluntarily to drop. As Commoner says, '. . . while a reduced death rate does, of course, increase the rate of population growth, it can also have the opposite effect – since families usually respond to a reduced rate of infant mortality by opting for fewer children . . . Similarly, although a rising population increases the demand on resources and thereby worsens the population problem, it also stimulates economic activity. This, in turn, improves education levels. As a result the average age at marriage tends to increase, culminating in a reduced birth rate . . .'

(c) 'What this means is that in order to bring the birth rates of the poor countries down to the low levels characteristic of the rich ones, the poor countries do not need to become as affluent . . . as the US. Achieving a per capita GNP only, let us say, one-fifth of that of the US – \$900 per capita per year – these countries could, according to the above relationship, reach birth rates almost as low as those of the European and North American countries . . . ' Commoner adds more statistics, showing that the world average per capita GNP is about \$803 per year – a level of affluence which is characteristic of a number of nations with birth rates of $\frac{20}{1{,}000}$. So, if the wealth of the world were evenly distributed among the people of the world, the entire world population should have a low birth rate. Commoner concludes with an evaluation of various proposed solutions, from family planning which puts 'the burden of remedying a fault created by a social and political evil – colonialism – voluntarily on the individual victims of the evil' to the 'life-boat ethic' whose astonishing proposals include: 'How can we help a foreign country to escape over population? Clearly the worst thing we can do is send food . . . Atomic bombs would be kinder. For a few moments the misery would be acute, but it would soon come to an end for most of the people, leaving a very few survivors to suffer thereafter.' His own solution is that 'the world population crisis, which is the ultimate outcome of the exploitation of poor nations by rich ones, ought to be remedied by returning to the poor countries enough of the wealth taken from them to give their peoples both the reason and the resources to voluntarily limit their own fertility. In sum, I believe that if the root cause of the world's population crisis is poverty, then to end it we must abolish poverty. And if the cause of poverty is grossly unequal distribution of the world's wealth, then to end poverty, and with it the population crisis, we must redistribute that wealth, among nations and within them.'

Review quiz

1(a) $\frac{2}{3} < \frac{5}{7}$
 (b) $8\frac{5}{4} > 4\frac{31}{56}$
2 100
3(a) 1975
 (b) 1976

16 The meaning of percentages

In 1976 in the USA, women held about 3,142,000 out of the 4,217,000 health-related jobs. At the same time, about 73,000 out of the 162,000 health administrators were women. Although these numbers clearly indicate that a great many health workers were women, these numbers make it difficult to see if women were proportionally represented in administrative jobs. In other words, did women hold equal portions of each health-related occupation, or did women hold the greater portion of the lower-paying jobs, while the fewer men in health-related jobs held the greater portion of the higher-paying, leadership jobs?

In order to use numbers to clarify this point, we convert the above data into percentages. This chapter shows you how to change comparison fractions into percentages. Then, the above numbers can be restated to show that whereas $\frac{3,142,000}{4,217,000}$ = 74.5 percent of all people in health-related jobs were women, only $\frac{73,000}{162,000}$ = 45.1% of the health administrators were women.

Per cent means 'per hundred' or 'out of 100'. A percentage is a comparison fraction, or ratio, whose denominator is 100. The above percentages indicate that about 75 out of every 100 health-related jobs were held by women, but only about 45 out of every 100 health administrators were women. Another way of stating this is that about $\frac{3}{4}$ ($\frac{75}{100} \div \frac{25}{25} = \frac{3}{4}$) of health-related jobs were held by women, but only about $\frac{1}{2}$ ($\frac{45}{100} \doteq \frac{50}{100} = \frac{1}{2}$ of the administrative jobs were held by women.

The percentage symbol, %, means divide by 100. For example, 75% means 75 ÷ 100, which can then be written as 0.75 (since 75 $\boxed{\div}$ 100 = 0.75) and 75% indicates $\frac{75}{100}$, or $\frac{3}{4}$, of a whole. Since all percentages are ratios with the same denominator, we can compare them directly. The larger the numerical value of the percentage, the larger the portion of the whole. 74.5% is a greater portion of a whole than 45.1%.

Finally, 100% represents the whole in percentages. This is consistent because 100% = 100 ÷ 100 = 1, and 1 represents the whole in fractions. The closer a percentage is to 100%, the larger the portion of the whole that it represents. For example, 98% represents almost the entire whole, whereas 3% represents a small portion of the whole.

SOLVED PROBLEMS

Drill

In problems 1–7 write each of the following percentages as a decimal.

Problem 1 47%

Solution Since the % symbol means divide by 100,
47% = 47 ÷ 100 = 0.47

Answer 47% = 0.47

Notes

1. It is important to be able to convert percentages to decimals or fractions so that when we solve problems with percentages we can multiply or divide with the value of the percentages. The percentage symbol is not a number that can be combined with other numbers. But 47% is just another way of writing 0.47, and 0.47 can be used in operations with other numbers.
2. One of my students, Dorothy Banuci, told me an easy way to remember that % means divide by 100: the % symbol looks like a slanted division symbol (÷) made up of a one and two zeros (for 100).
3. A short cut (a quick method which gives the correct answer) for dividing by 100 is to move the decimal point two places to the left.

   47 ÷ 100 = 47.0 ÷ 100 = 47.0 = 0.47

   When we focus on the meaning of division in Volume 2, you'll see why this short cut works.

You try this *(Answer at end of chapter)*

Write this percentage as a decimal
62%

Problem 2 30%

Solution 30% = 30 ÷ 100 = 0.30 = 0.3

Answer 30% = 0.3

You try this *(Answer at end of chapter)*

80%

Problem 3 5%

Solution $5\% = 5 \div 100 = 0.05$

Answer 0.05

You try this *(Answers at end of chapter)*
7%

Problem 4 45.1%

Solution Don't let more complicated-looking percentages freeze you. Whenever you remove the % symbol, you must divide the number part by 100:
$45.1\% = 45.1 \div 100 = 45.1 = 0.451$

Answer 0.451

You try this *(Answer at end of chapter)*
74.5%

Problem 5 $45\frac{1}{3}\%$

Solution $45\frac{1}{3}\% = 45\frac{1}{3} \div 100$

When working with percentages written in fraction form it is generally easier to change any fractions to decimals:

$45\frac{1}{3} \doteq 45.3$

$45\frac{1}{3}\% = 45\frac{1}{3} \div 100 \doteq 45.3 \div 100 = 0.453$

Answer 0.453
Note Use your judgement about rounding when changing fractions like $\frac{1}{3}$ to decimals. For most percentage problems, it is sufficient to round to the nearest tenth.

You try this *(Answer at end of chapter)*
$19\frac{1}{7}\%$

Problem 6 0.2%

Solution $0.2\% = 0.2 \div 100 = 0.002$

Answer 0.002

Notes 1 Often percentages smaller than 1% such as .2% will be written (as printed in this book) as 0.2% to make sure that the decimal point is emphasized in reading the number. (This is important in printing, where something as small as a point can often be missed.)

2 Intuitively, a percentage smaller than 1% simply means a portion smaller than 1%, or $\frac{1}{100}$, of the whole. In this case, 0.2 represents $\frac{2}{1,000}$ of the whole.

You try this *(Answer at end of chapter)*
0.9%

Problem 7 230%

Solution 230% = 230 ÷ 100 = 2.30 = 2.3

Answer 2.3

Note Percentages larger than 100% generally occur in problems involving percentage change. They will be treated in Volume 2.

You try this *(Answer at end of chapter)*
465%

Problem 8 Write 8.3% as a fraction.

Solution The simplest way to write any percentage as a fraction is first to write the percentage as a decimal and then use the rules in Chapter 11 to write the decimal as a fraction.

8.3% = 8.3 ÷ 100 = 0.083 = $\frac{83}{1,000}$

Answer $\frac{83}{1,000}$

Note If not otherwise indicated, use your judgement about whether or not to reduce the fraction.

You try this *(Answer at end of chapter)*
17.2%

In problems 9–11 write each of the decimals as percentages.

Problem 9 0.82

Solution Changing decimals to percentages is the opposite of changing percentages to decimals. Here we multiply by 100 and add the % symbol:

0.82 = 0.82 × 100% = 82%

Answer 82%

Notes 1 A short cut for multiplying by 100 is to move the decimal point two places to the right.

0.82 × 100 = 0.82 = 82.0 = 82%

2 This procedure makes sense since if we were to change back to a decimal, 82% = 82 ÷ 100 = 0.82, we get the decimal with which we started.

You try this *(Answer at end of chapter)*
0.37

Problem 10 0.067

Solution 0.067 = 0.067 × 100% = 6.7%

Answer 6.7%

You try this *(Answer at end of chapter)*
0.088

Problem 11 4.52

Solution 4.52 = 4.52 × 100% = 452%

Answer 452%

You try this *(Answer at end of chapter)*
6.29

Problem 12 Write $8\frac{3}{4}$ as a percentage.

Solution The simplest way to write any fraction as a percentage is first to write the fraction as a decimal using the rules in Chapter 14 and then write the decimal as a percentage.

$\frac{3}{8} = 3 \div 8 = 0.375$

$2\frac{3}{8} = 2.375 = 2.375 \times 100\% = 237.5\%$

Answer 237.5%

You try this *(Answer at end of chapter)*
$4\frac{5}{8}$

Applications

Problem 13 In the USA in 1976, 15.5% of all health-related jobs were held by blacks and other people of colour. In 1976, what percentage of all health-related jobs were held by whites?

Solution First, we must determine what the whole is and how it is being broken down. In this case, the whole is all health-related jobs. These jobs are being broken down according to whether they are held by a black or other people of colour or by a white. Since this covers all the possibilities and the whole is 100%, we can use subtraction:

100% − 15.5% = 84.5%

Answer In 1976, 84.5% of all health-related jobs were held by whites.

Notes

1. If we had been given that 15.5% of all health-related jobs were held by blacks, we could not determine the percentage of such jobs held by whites. In this case the whole is made up of three parts – blacks, whites, and other people of colour. Since we would know only one part, we couldn't find the other missing parts.
2. Here, as in Chapters 14 and 15, you need to perform some basic operations (in this case, subtraction) in order to get an answer. Because this text presents the meanings of various numerical and algebraic maths concepts before focusing on the operations, the maths learning is not linear. So, you'll be doing some basic operations before fully understanding the meaning of the operations. If this is confusing to you now, try to tolerate that ambiguity – it will become clear as you work through Volume 2. All real learning involves confusion with new material, questioning and reviewing and understanding past learning.

You try this *(Answer at end of chapter)*

In 1976, 2.8% of dentists were black or other people of colour. In 1976, what percentage of dentists were white?

UNSOLVED PROBLEMS

(Answers at end of chapter)

Drill

1 Write each of the following percentages as a decimal.

(a) 23% (e) 8.9%
(b) 60% (f) 32½%
(c) 2% (g) 100%
(d) 62.3% (h) 250%

2 Write each of the following percentages as a fraction.

(a) 75% (d) 235%
(b) 3% (e) $10\frac{1}{2}$%
(c) 3.5% (f) 300%

3 Write each of the following decimals as a percentage.

(a) 0.62 (e) 0.405
(b) 0.70 (f) 6
(c) 0.07 (g) 0.6
(d) 0.7 (h) 4.6

4 Write each of the following fractions as a percentage.

(a) $\frac{42}{100}$ (d) $\frac{1}{4}$
(b) $\frac{8}{10}$ (e) $\frac{2}{9}$
(c) $\frac{4}{5}$ (f) $2\frac{7}{8}$

Thought

5 Find the error in each of the following groups of conversion problems. Then do the problems correctly.

(a) Write each percentage as a decimal:
$10\frac{1}{2}\% = 10.5$
$6\frac{2}{3}\% = 6.7$
$7\frac{9}{10}\% = 7.9$

(b) Write each decimal as a percentage:
$0.5 = 5\%$
$0.2 = 2\%$
$0.9 = 9\%$

6 Read the following letter to *The Mathematics Teacher* (February 1982).

(a) Summarize Waldron's main point in your own words.
(b) Do you agree or disagree, and why?

'Alphabets and percent

'Even though we stress that percent means hundredths, some students have difficulty writing decimals as percents. To help them recall the correct procedure, I ask them to write down the letters of the alphabet and notice that d comes before p. Since the alphabet goes from left to right, from d to p, then the decimal point should be moved to the right to find its location as a percent. For example:

d	*p*
0.167	16.7%
2.0	200%
0.02	2%

'John Waldron
Mario Umana Harbor School of
Science and Technology
East Boston, MA 02171'

7 Briefly discuss the humour in the following cartoon (*In These Times*, 19 February 1986) and what knowledge of maths concepts you need to understand the joke.

8(a) Create and solve a quiz which reviews the major concepts of this chapter.

(b) Write about why you chose the problems you did to review this chapter.

Applications

9 Describe the conclusions you can reach from the data in the following chart (from *New Society* Database, 4 March 1988).

Percentage of smokers by occupation in Britain

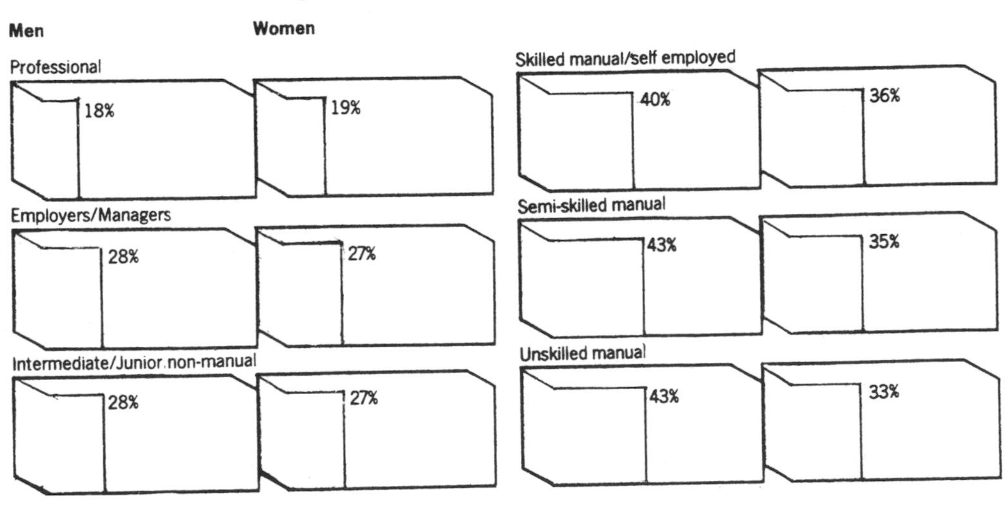

10 Use the table below (from the *Statistical Abstracts of the United States 1977*) to solve the following problems:

No. 146. Employed persons in selected health occupations: 1972 to 1976: persons sixteen years old and over (annual averages)

Occupation	1972 (1,000)	1975 (1,000)	1976 Number (1,000)	1976 Percent Females	1976 Percent Black and other
Health occupations	**3,511**	**4,041**	**4,217**	**74.5**	**15.5**
Physicians, dentists, and related	624	647	671	10.9	7.2
Dentists	107	110	107	1.9	2.8
Pharmacists	126	119	123	16.3	6.5
Physicians, medical and osteopathic	328	354	368	12.8	9.5
Other practitioners	64	64	75	5.3	1.3
Nurses, dietitians, and therapists	949	1,126	1,203	92.6	10.8
Registered nurses	801	935	999	96.6	10.8
Dietitians and therapists	148	192	204	72.5	10.8
Health administrators	118	152	163	45.1	6.8
Health technologists and technicians	315	397	436	72.9	13.1
Clinical laboratory	142	177	192	74.5	16.7
Radiology	68	79	80	68.8	10.0
Other	104	142	164	73.2	9.1
Health service	1,505	1,718	1,745	89.6	23.4
Health aides and trainees, except nursing	148	219	229	83.8	21.0
Nursing aides, orderlies, and attendants	912	1,001	1,002	86.8	26.4
Practical nurses	342	370	381	97.4	23.1
Other	103	128	134	97.0	4.5

Source: US Bureau of Labor Statistics, *Employment and Earnings*, October 1973, January 1976, and January 1977

(a) Briefly describe the kinds of information contained in this table.
(b) In 1976, what percentage of dentists were men?
(c) In 1976, what percentage of registered nurses were white?
(d) In 1976, what percentage of practical nurses were men?
(e) In 1976, what percentage of clinical lab technicians were white?
(f) In 1976, which health-related occupation had the smallest percentage of women workers?
(g) In 1976, which health-related occupation had the smallest percentage of white workers?
(h) How many registered nurses were there in 1975?
(i) Create and solve a maths problem based on the information in this chart.
(j) What conclusions might be drawn from the information contained in the chart?
(k) What additional conclusions might be drawn from the information contained in this updated table from the *Statistical Abstracts of the United States 1986*?

No. 161 Employed persons in selected health occupations: 1984: civilians sixteen years old and over (annual averages)

Occupation	Number (1,000)	Percent Female	Percent Black
Total	5,658	76.6	13.1
Health diagnosing occupations*	775	13.4	3.8
Physicians†	520	16.0	5.0
Dentists	138	6.2	0.9
Health assessment and treating occupations	1,929	86.4	7.1
Registered nurses	1,402	96.0	7.6
Pharmacists	162	28.5	2.9
Dietitians	70	95.1	16.1
Therapists	251	76.8	5.4
Physicians' assistants	43	37.2	3.3
Managers, medicine and health	96	61.2	6.4
Health technologists and technicians*	1,112	83.4	13.1
Clinical laboratory	281	75.6	12.8
Dental hygienists	62	99.0	2.6
Radiology	110	66.9	7.4

continued

No. 161 *(continued)*

Occupation	Number (1,000)	Percent Female	Percent Black
Licensed practical nurses	416	96.3	17.1
Health service occupations	1,746	90.3	24.2
Dental assistants	170	98.2	4.4
Health aides, except nursing	341	86.4	16.8
Nursing aides, orderlies, and attendants	1,235	90.4	29.0

* Includes other occupations not shown separately.
† Medical and osteopathic.

Source: US Bureau of Labor Statistics, *Employment and Earnings*, January 1985.

11 Use the following information from *Diet for a Small Planet* (Lappé, 1975) to solve the problems below.

Percentage of daily protein allowance in an average serving

Food item	Male (154 lb)	Female (128 lb)
	%	%
Cottage cheese, uncreamed, 3½ oz	30	36
Yoghurt from skim milk, 1 cup	16	20
Cheddar cheese, 1 oz	12	14
Soya beans, ¼–⅓ cup, dry	23	28
Chick-peas (garbanzos), ¼–⅓ cup, dry	12	14
Tofu, 3½ oz	12	14
Pumpkin seeds, 2 tbsp	12	14
Peanut butter, 2 tbsp	7	8
Spaghetti, 1 cup, cooked	7	8
Brown rice, ⅓ cup	7	8
Whole-wheat bread, 1 slice	2	3
Collards, ½ cup, cooked	5	6
Mushrooms, 10 small or 4 large	5	6
Mustard greens, ½ cup, cooked	3	4
Wheat germ, 2 level tbsp	5	6

(a) Why do you think the percentages are different for men and women?

(b) Find two different, realistic combinations using the above food items that would provide a complete daily allowance of protein for a 154-lb male.

(c) Find two different, realistic combinations using the above food items that would provide a complete daily allowance of protein for a 128-lb female.

(d) What other factors, besides protein, are important considerations in eating a healthful diet?

(e) Read Part II of *Diet for a Small Planet*, 'Bringing Protein Theory Down to Earth'. Briefly explain what complementing proteins means and why it increases the percentage of daily protein allowance.

12 Summarize the main point of the following graph (from *New Society* Database, 23 January 1987):

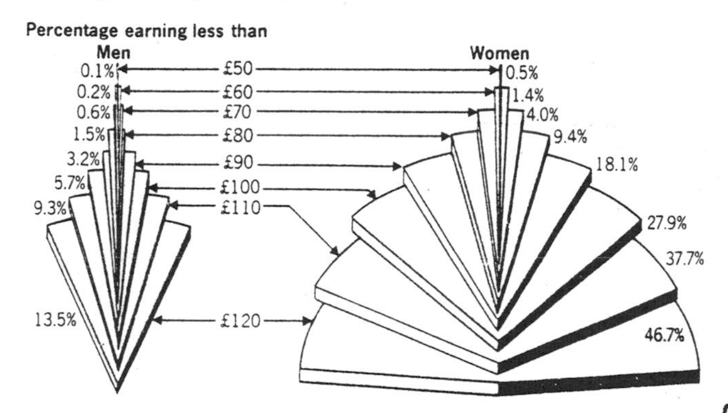

The low paid in Britain *Source: New Earnings Survey*, 1986

13 Briefly summarize the '4% Deception' in your own words (*Nutrition Action*, July/August 1986).

'The 4% deception

'Jeepers! Milk has less than four percent fat!" exclaims Betty Boop, her peepers open wide. Boop's surprise is part of the Middle Atlantic Milk Marketing Association's new advertising campaign to convince consumers that even whole milk is low in fat.

'Because water accounts for 87 percent of whole milk's weight, the *percentage* of weight contributed by fat is indeed quite small. But that doesn't diminish the actual *amount* of fat in whole milk. Each glass still contains roughly two pats' worth of butterfat. Half of whole milk's calories come from fat.

'In a complaint to the Maryland State Attorney-General, the non-profit Baltimore Vegetarians charged that the ad is misleading and harmful to consumers. CSPI agrees. Jeepers, Betty! Did you know that whole milk is the second largest source of saturated fat in the average American's diet!'

14 Read 'Color on the Court' by Jerome Karabel and David Karen (*In These Times*, 10 February 1982).
(a) What is the main idea and how do the numbers support that main idea?
(b) Do you agree with the authors' arguments and why or why not?

'Color on the court

'The National Basketball Association was founded in 1949 as an all-white enclave – two years *after* Jackie Robinson broke the color line in major league baseball. Professional basketball has come a long way since then, with quality of performance on the court gradually casting aside considerations of skin color. In the 25 years between 1955 and 1980, the proportion of blacks in the league has gone from less than 10 percent to more than 70 percent.

'Blacks now set the tone in the NBA; the slower, more deliberate game of the '50s has been replaced by a running and jumping brand of basketball exemplified not by the two-handed set shot but by the slam dunk. As recently as 1958, an all-white St Louis Hawks team won the NBA; in today's NBA, however, a team can simply not succeed without drawing on the talents of black players.

'Yet if the players are free of many of the racial prejudices that pervade American society, the fans most assuredly are not. And the owners, whatever their private feelings on racial matters, cannot afford to ignore this.

'Ted Stepien, current owner of the Cleveland Cavaliers, addressed the issue of race with a crude bluntness shortly before he purchased the team: "This is not to sound prejudiced, but half the squad would be white . . . White people have to have white heroes. I myself can't equate to black heroes. I'll be truthful – I respect them, but I need white people. It's in me. And I think the Cavs have too many blacks, 10 of 11. You need a blend of black and white. I think that draws, and I think that's a better team."

'While Stepien's personal views on race may be unusual among NBA owners, there may be a kernel of truth in what he is saying. Many white fans may, in fact, be unwilling to give their support to an overwhelmingly black team, as even a cursory glance at the current attendance figures for the Larry Bird-led Boston Celtics versus the figures for the Celtics during the '60s, when most of their important players – including the great Bill Russell – were black, would attest. And indeed, when asked about his remarks, Stepien declared that they were issued "in a context of marketing". (Perhaps by sheer coincidence, the overwhelmingly black Cavaliers team of 1979–80 was transformed, in the year after Stepien's purchase, to one composed of six whites and five blacks.)

'Stealin'

'That race may be a consideration in professional basketball today is hardly news to black players. Indeed, according to David Halberstam's superb new book on the NBA, *The Breaks of the Game* . . ., there is a widespread feeling among black players that their jobs are less secure, especially those of bench players. Many teams, Halberstam contends, fill their lower bench positions with marginal white players as a means of pacifying the fans. Not surprisingly, black players resent this and, according to Halberstam, even have a name for it: "stealin'".

'Anecdotes of teams allowing racial considerations to enter into personnel decisions abound. There was the time, for instance, that Joe Caldwell, averaging nearly 20 points a game, was sent to the bench by an Atlanta team desperate for a starting white.

'But are these anecdotes simply isolated instances, or is there an underlying racial pattern in the NBA? To answer this question, we gathered data for the 1980–1 season for each of the NBA's 23 teams on: the race of players; the racial composition of the team's geographical area; and (in order to determine which players spent the most time on the bench) the number of minutes logged by each player.

'We first checked for a relationship between where white players were concentrated and the racial composition of the league's cities. If race remains a factor, then white players should be disproportionately concentrated on teams in the league's whitest cities. To test this, we divided the league's cities into three categories: less than 10 percent black, 10 to 20 percent black, and more than 20 percent black.

'The results, shown in the accompanying table, reveal a clear pattern: the whiter the category of city, the higher the proportion of white players. The probability that this relationship could have occurred by sheer chance is less than 1 in 100. Ted Stepien, it seems, may not be alone in thinking that stocking a team with white players is a sound business practice.

'Do fan-pressured NBA teams fill their benches, as Halberstam suggests, with marginal white players? To determine whether they do, we divided the players on every team into two categories: the "starters", the five players who play the most minutes per game and the "reserves", the remaining six players on the team. Our findings indicate that – as predicted by those who claim that marginal white players receive special treatment – whites are indeed underrepresented among the starters and overrepresented among the reserves: 22.6 percent versus 31.2 percent. But this relationship is not a strong one; a pattern of this sort could have occurred by pure chance approximately one time in seven.

'To further test the "marginal white player" hypothesis, we divided the reserves into two groups: those occupying positions six through eight, measured in minutes on the court, and those occupying positions nine through eleven. If marginal white players were kept on simply as a bone to the fans, we reasoned, then they ought to be overrepresented toward the back of the bench. Our results, however, revealed that this was *not* the case; whereas 37.7 percent of the players in positions six to eight were white, only 24.6 percent of the players in positions nine to eleven were white.

'The popular image of the rarely used white player sitting at the far end of the bench does nonetheless receive some support from our data. In those eight cities with a black population of less than 10 percent – the very cities that would presumably be most subject to white fan pressure – the percentage of white players toward the back of the bench (positions nine to eleven) was 37.5 percent compared with *zero* for the league's five blackest cities.

'But the limitations of our study prevented us from looking at all the ways that race might affect decisions by NBA owners, general managers and coaches. We had no way of testing whether racial factors influence who gets into the NBA – if it is possible that talented black players are occasionally cut in favor of whites of lesser ability. Racial considerations could also enter into decisions about who gets to play, especially in games that are out of reach anyway. Our study may therefore underestimate the role of race in today's NBA.

'Our findings, though hardly definitive, are more than sufficient to establish that the NBA is anything but color blind. The league has progressed dramatically from the days in 1960 when the fine black guard of the St Louis Hawks, Sihugo Green, found himself under orders to rebound and play defense, leaving the scoring – insofar as was possible – to the whites. But professional basketball is still a long way from meritocracy. And the economics of the box office suggest that the path to true color blindness may be a tortuous one indeed.'

Racial composition of cities*	Racial composition of teams (%)		
	Black	White	Total (%)
Less than 10% black: Boston, Denver, Phoenix, Portland, San Antonio, San Diego, Seattle, Utah	63.6	36.4	100 (88)†
From 10% to 20% black: Dallas, Golden State, Indiana, Kansas City, Los Angeles, Milwaukee, Cleveland, Houston, New Jersey, Philadelphia	72.7	27.3	100 (110)
More than 20% black: Atlanta, Chicago, Detroit, New York, Washington	87.3	12.7	100 (55)
Total (League as a whole)	72.7 (184)	27.3 (69)	100 (253)

* City population figures based on 1980 Census Bureau data for the relevant Standard Metropolitan Statistical Area.
† Numbers in parentheses represent the actual number of players.

15 Use the following facts (taken from Neft *et al.*, *The World Book of Odds*), and the given formula, to solve the problems below. Given that the odds of a particular event occurring are A to B, the percentage chance that the event will occur is obtained by writing the fraction A ÷ (A + B) as a percentage. For example, the odds that a scheduled passenger US airline flight will be at least 15 minutes late are 1 to 4. The percentage chance that a scheduled passenger US airline flight will be at least 15 minutes late is

$\frac{1}{1}+4 = \frac{1}{5} = 0.2 = 20\%$.

(a) The odds that a published fiction book in the USA will become a best seller are 1 to 50. Find the percentage chance that such a book will become a best seller.
(b) The odds that a published non-fiction book in the USA will become a best seller are 1 to 200. Find the percentage chance that such a book will become a best seller.
(c) In the USA the odds that a couple whose marriage ended in divorce or annulment had one or more children are 11 to 8. Find the percentage chance that such a couple had one or more children.
(d) The odds in favour of drawing an X on the first draw of seven

Scrabble tiles are 1 to 13. Find the percentage chance of drawing an X on the first draw of seven Scrabble tiles.

(e) The odds of making a flush in five-card draw poker (with no wild cards) by drawing one card to a four-card flush are 1 to 4.2. Find the percentage chance of making such a flush.

(f) The odds of making a flush in five-card draw poker (with no wild cards) by drawing two cards to a three-card flush are 1 to 23. Find the percentage chance of making such a flush.

(g) The odds of finding a pearl in a pearl oyster are 1 to 12,000. What is the percentage chance of finding a pearl in a pearl oyster?

REVIEW QUIZ (the meaning of percentages) (*Answers at end of chapter*)

1 Write 6.8% as a decimal.

2 Write 150% as a fraction.

3 Write 0.009 as a percentage.

4 Write $\frac{3}{11}$ as a percentage.

5 In 1976 in the USA, 97.4% of all practical nurses were women. What percentage of all practical nurses were men?

ANSWERS: CHAPTER 16

Solved problems ('You try these')

1 0.62
2 0.8
3 0.07
4 0.745
5 $19\frac{1}{7}\% = 19.1\% = 0.191$
6 0.009
7 4.65
8 $17.2\% = 0.172 = \frac{172}{1,000}$
9 37%
10 8.8%
11 629%
12 $4\frac{5}{8} = 4.625 = 462.5\%$
13 $100\% - 2.8\% = 97.2\%$

Unsolved problems

1(a) 0.23
(b) 0.6
(c) 0.02
(d) 0.623
(e) 0.089
(f) $32\frac{1}{2}\% = 32.5\% = 0.325$
(g) $100\% = 1.00 = 1$
(h) $250\% = 2.5$
2(a) $\frac{75}{100}$ (which can be reduced to $\frac{3}{4}$)
(b) $\frac{3}{100}$
(c) $3.5\% = 3.5 \div 100 = 0.035 = \frac{35}{1,000}$
(d) $235\% = 2.35 = 2\frac{35}{100}$
(e) $10\frac{1}{2}\% = 10.5\% = 0.105 = \frac{105}{1,000}$
(f) $3 = \frac{3}{1}$
3(a) 62%
(b) 70%
(c) 7%
(d) 70%
(e) 40.5%
(f) 600%
(g) 60%
(h) 460%
4(a) 42%
(b) 80%
(c) $\frac{4}{5} = 0.8 = 80\%$
(d) 25%
(e) $\frac{2}{9} \doteq 0.22 = 22\%$
(f) $2\frac{7}{8} = 2.875 = 287.5\%$
5(a) Here the person changed the number part of the percentage to a decimal correctly, but just dropped the percentage sign without dividing by 100. Correctly:
$10\frac{1}{2}\% = 10.5\% = 0.105$
$6\frac{2}{3}\% \doteq 6.7\% = 0.067$
$7\frac{9}{10}\% = 7.9\% = 0.079$

(b) Here the person multiplied by 10% instead of 100%. Correctly:
0.5 = 0.50 = 50%; 0.2 = 20%; 0.9 = 90%

6 To remember the short cut that, to multiply by 100, we move the decimal point two places *to the right*, remember that in *d*ecimals to *p*ercentages, to get from 'd' to 'p' in the alphabet you go *to the right*.

7 This is a joke about a rigged election, rigged because 312% > 100% and if each person votes once, the most you can have is 100% of the vote., But if the election is rigged, if people stuff the ballot boxes voting more than once, then you'll have *more* votes than voters, so you could have more than 100% of the votes. Understanding of this cartoon is also enhanced with the background knowledge that Marcos, the former dictator of the Philippines, did rig his elections.

9 According to *New Society*,

'Smoking is linked to occupation. For men, the prevalence of smoking increases the lower the socioeconomic group. While 18 per cent of professional men smoke, 28 per cent of intermediate and junior white collar workers and 43 per cent of unskilled manual workers do so. The picture is somewhat different for women, with smoking most common among skilled manual and self-employed non-professionals. Professionals are the only occupational group in which a higher proportion of women smoke than men.'

10(a) This matrix chart breaks down the various health occupations comparing: the total number of people in each occupation over the years 1972, 1975, and 1976; the percentages of men *v.* women; and white *v.* black and other people of color in 1976. (You get the percentage of the group not listed by subtracting the listed group from 100%.)

(b) 100% − 1.9% = 98.1%.

(c) 100% − 10.8% = 89.2%.

(d) 100% − 97.4% = 2.6%.

(e) 100% − 16.7% = 83.3%.

(f) Dentists.

(g) Nursing aides, orderlies, and attendants.

(h) The (1,000) under the 1975 in the table means the numbers in the table are rounded to the nearest thousands, and to save space the three zeros are left off. So the number of registered nurses in 1975 was not 935 – but 935,000. The total number in health occupations in 1975 was 4,041 thousand or 4,041,000.

(j) Most health workers in 1976 were women; these women were mainly nurses and technicians; the more 'prestigious', high-paying jobs of doctors, etc. were held by white men; also, most men and women of color worked in the less prestigious, lower-paying health jobs.

(k) The conclusions are the same. Direct comparisons are made difficult because the 1976 'Black and other' data are only given for 'Black' in 1984. So, for example, we can't tell if the portion of black dentists has actually dropped or if the greater 1976 portion included Asian and native American dentists.

11(a) Men weigh, on average, more than women, so the same amount of any nutrient will make up a smaller portion, or percentage, of their protein requirements.

(b) One is the cottage cheese and 2 cups of yoghurt and soya beans and tofu and pumpkin seeds (30% + 32% + 23% + 12% + 12% = 109%). You don't have to have exactly 100%, just *at least* 100%.

(c) One is the cottage cheese and soya beans and 7 oz tofu and brown rice (36% + 28% + 28% + 8% = 100%).

12 A greater percentage of women than men earn less than £50, £60 . . . £120. For example 46.7%, or almost half of all women, earn less than £120, whereas only

13.5% of all men earn less than that amount; that is, 86.5% of all men earn more than £120.

13 This is a case where the percentage is misleading, and the actual amount is the relevant figure. (In the examples above about, say, women in politics, the percentage reveals more than the amount, showing that most politicians in each of the groups were men – just the number of, say, women governors would not make it clear that only 4% were women.) The *portion* of weight fat is small because milk is mainly water, but there is still a large *amount* of fat in milk. The portion, or percentage of calories is also another, not misleading figure that could have been presented: half, or 50%, of whole milk's calories come from fat.

14 Although from 1955 to 1980 the proportion of blacks in the NBA has gone from less than 10% to more than 70%, race is still a factor in the composition of teams. The numerical data that exist support this: the 'whiter' the city, the higher the percentage of white players on the team (the probability that this could occur by sheer chance is less than 1%); in the 'whitest' cities, the percentage of 'marginal white players' on the bench was 37.5% compared with 0% for the NBA's 'blackest' cities. Also, this study's limitations prevented a look at determining if race is a factor in, for example, who gets to play. Note that the statistical analysis done on the numbers is important because it shows that the anecdotes about racism in the NBA are not simply isolated instances, but part of a pattern.

15(a) $\frac{1}{1+50} = \frac{1}{51} \doteq 0.02 = 2\%$

(b) $\frac{1}{1+200} = \frac{1}{201} \doteq 0.005 = 0.5\%$ (note we had to round to the nearest thousandth because there were no digits in the tenths and hundredths places).

(c) $\frac{11}{11+8} = \frac{11}{19} \doteq 0.579 = 57.9\%$

(d) $\frac{1}{1+13} = \frac{1}{14} \doteq 0.071 = 7.1\%$

(e) $\frac{1}{1+4.2} = \frac{1}{5.2} = 0.192 = 19.2\%$

(f) $\frac{1}{1+23} = \frac{1}{24} \doteq 0.042 = 4.2\%$

(g) $\frac{1}{1+12{,}000} = \frac{1}{12{,}001} \doteq 0.00008 = 0.008\%$

Review quiz

1 0.068

2 $1\frac{50}{100}$ or $1\frac{1}{2}$

3 0.9%

4 27.3%

5 2.6%

17 The meaning of measurements

A measure is a numerical description of something. Your height, your weight, your age, and your temperature are all measures of some aspect of you. Batting averages, field goal shooting percentages, and average yards per carry are all measures of some aspect of particular sports (baseball, basketball, and American football, respectively). Our units of measurement are uniform, or *standard*, because each unit represents the same quantity to everyone, everywhere. When we want to measure something, we:

1. Choose an aspect about it that is measurable (such as its height or its speed).
2. Choose an appropriate unit of measurement (such as feet or kilometres per hour).
3. Choose an appropriate measurement instrument (such as a ruler or a speedometer).
4. Determine how many of the units are 'contained in' or 'make up' the aspect you are measuring.

SOLVED PROBLEMS

Thought Problem 1 What are some reasons why the measures we assign to things might not be exact?

Solution There are two main reasons why measures might not be exact: human 'limitations' and the nature of measurement instruments. For example, it is virtually impossible to determine the exact number of people living in a specific city – people move every day, new babies are born every hour, the census taker might accidentally (or not so accidentally) miss a block, and so on. Because of these human 'limitations', many measures are only approximate.

In Chapter 10 you were introduced to the ruler. The smaller the divisions of a ruler, the more accurately it can measure the length of an object. Because we can (in theory) keep making the divisions smaller, we can always (in theory) get a more accurate measurement. For example,

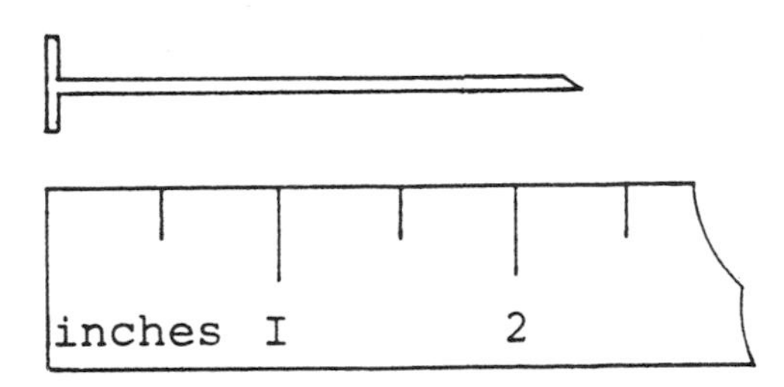

measuring the above nail with a ruler whose divisions occur every ½ inch, we say the nail is 2½ inches long. It is somewhat less than that, but we can't tell exactly how much shorter. No matter how small the divisions of a ruler, there will always be some error involved. (Even the line marking the ruler divisions has a certain thickness: where does an object measure exactly 2 inches – if it coincides with the beginning of the 2-inch mark or the end of the 2-inch mark?) So, reading a measurement is a rounding procedure – you are determining which ruler division is closest to the length of the object. Here, we are rounding to the nearest half inch. We mentally halve the length between the 2-inch and the $2\frac{1}{2}$-inch marks on the ruler and determine that the nail ends in the part closer to $2\frac{1}{2}$ inches.

You try this *(Answer at end of chapter)*

Give an example of a measure which is exact.

UNSOLVED PROBLEMS *(Answers at end of chapter)*

Thought

Problems 1–12 are adapted from Bell, 1972.

1(a) List as many measures that apply to your environment as you can, such as your height, your expenses, the amount of time you spend studying, and so on.

(b) For each measure, list the appropriate unit.

2 In many cases, units of measurement involve more than one measure. For example, speed is measured in *miles* per *hour* and

baseball batting averages are measured in *number of hits* per *number of times at bat*. List as many of these 'compound' measures as you can.

3 Visit three different places of work which use measures, such as an airport, a weather bureau, and a construction company. List the types of measures used at each place.

4(a) List as many measuring instruments as you can.
(b) What does each instrument measure?
(c) In what units does each instrument measure?

5(a) Decide which of the following items can be measured exactly and which can only be measured approximately:
- The number of siblings you have.
- Your height.
- The amount of money in your pocket.
- The amount of money in your bank account.
- Your friend's age.
- The distance from your house to your school.
- The distance from New York to Boston.

(b) List two other items that can be measured exactly.
(c) List two other items that can only be measured approximately.

6 List as many aspects of the world as you can that cannot be measured.

7 The origin of some of our common measures gives us a good understanding of why we standardize measures. The following sketch shows some common measures based on dimensions of the human body.

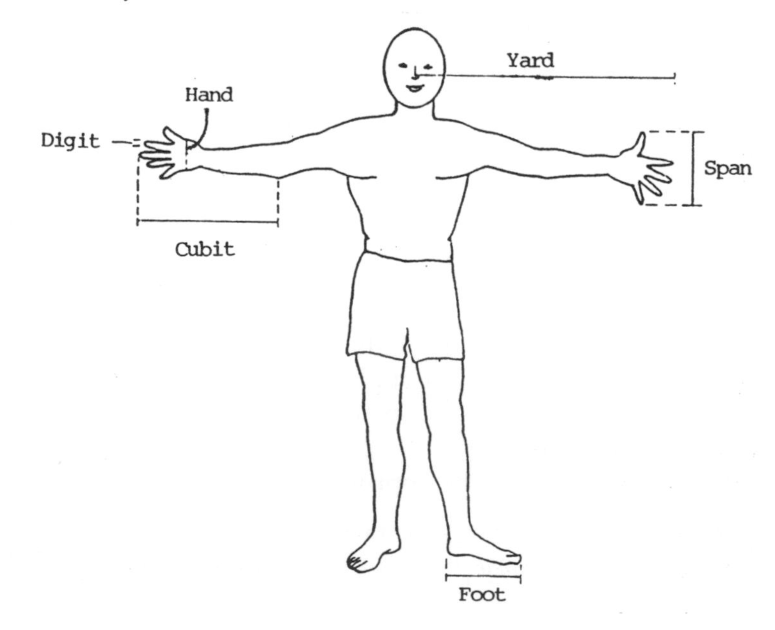

The 'yard', taken as the distance from the tip of the nose to the tips of the fingers, was often used in stores for measuring cloth.

(a) With a string, see how long this length is with a number of people of different ages.

(b) Why do you think we standardize the measure of a yard?

8 A span is the distance from the end of the little finger to the end of the thumb in an outstretched hand.

(a) Find the length of your span.

(b) Use your span to measure five things. (List the measurement of each thing.)

9 A pace is the distance between the tip of one foot to the tip of the same foot when it next touches the ground after a full step has been taken.

(a) Find the length of your pace by walking ten full steps, measuring the distance, and then dividing by 10.

(b) Why does the above method give you the length of your pace?

(c) What is the longest pace you can walk?

10 When the Lilliputians measure Gulliver for clothes in *Gulliver's Travels*, they measure his thumb and then apply a rule that 'twice around the thumb is once around the wrist, and so on to the neck and waist . . . ' Check this rule on yourself and on some friends.

11 Some common measures that can be used to get approximations are:

– lines in US pavements are frequently spaced 5 ft apart;
– a brick is about 8 in. long;
– a compact car is about 15 ft long;
– most cars are about 6 ft wide;
– an American city block is often $\frac{1}{8}$ mile long.

(a) Find two other such common measures.

(b) Use three of the above 'units' to find some approximate measures. For example, approximate the length of a building using the brick measure.

12(a) Invent your own unit of measure and some smaller and larger related units. (For example, if the foot corresponded to your unit, the inch would be a related smaller unit (12 in. = 1 ft), and the yard would be a related larger unit (1 yd = 3 ft).)

(b) Measure at least four items with your units of measure.

13 Use the following article (*Scientific American*, January 1979) to answer the questions below.

'Metrication at the crossroads

'In spite of its name the Metric Conversion Act of 1975 does not provide for the adoption of the metric system as the predominant system of measurement in the US. The act merely establishes a Federal agency to coordinate and support the efforts of those in business, Government and education who choose to convert to the metric system. According to a recent study conducted by the General Accounting Office, the fact that conversion is voluntary rather than mandatory means that many of the presumed benefits of metrication may never be realized.

'Proponents of metrication have always maintained that the metric system is easier to work with because the measuring units come in multiples of 10. In a familiar example the liter, the metric unit of volume, consists of 1,000 milliliters, whereas a quart consists of a clumsy 32 fluid ounces. It would therefore appear that the metric system would make it easier to compare the unit prices of variously packaged items in a supermarket. Such is the case, however, only when the actual dimensions of products are changed to metric dimensions ("hard conversion"), not merely when the customary measurement units are replaced by their metric equivalents ("soft conversion"). In other words, little is accomplished if one quart of a beverage is simply relabeled 0.95 liter; rather, the size of the container must be changed to one liter, which happens to equal 1.06 quarts. Only in the latter case will the ease of working with multiples of 10 come into play. The goal of hard conversion, the General Accounting Office maintains, may well be unattainable without the imposition of laws and regulations.

'Even in industries that have undertaken hard conversion, price comparisons are not necessarily easier to make. The distilled-spirits industry is a case in point. Before the industry began to adopt the metric system 94 percent of its sales were in five customary sizes: $\frac{1}{2}$ gallon, quart, $\frac{4}{5}$ quart, pint and $\frac{1}{2}$ pint. Except for the $\frac{4}{5}$ quart these sizes are all integral multiples of one another, enabling consumers to make price comparisons easily. The 200-, 500- and 750-milliliter and the one- and 1.75-liter containers that have come to replace the customary sizes are more difficult to compare in price.

'Curiously the average unit prices of the 200-milliliter and 1.75-liter sizes, the ones whose prices are the most difficult to compare, showed the greatest increase over their nonmetric equivalents. The unit price of the 200-milliliter size is 11.4 percent higher than its $\frac{1}{2}$-pint equivalent and the unit price of the 1.75-liter size is 6.1 percent higher than its $\frac{1}{2}$-gallon equivalent. The General Accounting Office does not predict whether or not these price hikes will persist once the industry is completely metricated. The report does indicate, however, that the metrication of the entire US would cost billions of dollars.

'The General Accounting Office study calls on Congress to commit itself either to the traditional US system or to the metric system but not to both. The present course of voluntary metrication will lead to a dual system that is "impractical, inefficient, uneconomical and confusing". The General Accounting Office does not say, however, which of the two systems it prefers, although it belittles many of the presumed advantages of metrication. The US is now the only major nation not using the metric system, and it may still be years before Americans join the rest of the world in realizing that 28.3 grams of prevention are worth 0.453 kilogram of cure.'

(a) What is the main idea of this article?
(b) List all the reasons for converting to the metric system discussed in this article.
(c) List all the reasons against converting to the metric system discussed in this article.
(d) Write a brief paragraph supporting your opinion about whether or not the United States should convert to the metric system.
(e) Explain what 'Except for the $\frac{4}{5}$ quart these sizes are all integral multiples of one another . . . ' means. (See middle of third paragraph.)
(f) Explain why it is easy to make price comparisons when sizes are integral multiples of one another.
(g) Why are the 200 ml, 500 ml, 750 ml, 1 litre, and 1.75 litre containers difficult to compare in price?
(h) Explain the humour in the last sentence.

14 Read '299,792,458 Meters per second' (*Scientific American*, June 1982).
(a) What is the main point?
(b) About how many meters per second will the speed of light be defined as?
(c) Write the previous definition of a meter as a fraction and as a decimal fraction.

'299,792,458 Meters per second

'The special theory of relativity states that the speed of light in a vacuum is a constant of nature. Measurements of the speed made in any frame of reference yield the same value. The speed of light may soon become a constant of another kind: a defined quantity rather than a measured one, with a value fixed by international convention. The value likely to be adopted is 299,792,458 meters per second.

'The change in status would result from a proposed redefinition of the meter. Originally the meter was set equal to one ten-millionth of the distance from the North Pole to the Equator, measured on the meridian passing through Paris; somewhat later it was defined by reference to a platinum-iridium bar; since 1960 the standard meter has been a certain multiple of the wavelength of light emitted by krypton gas. In the new definition the standard unit of length would no longer be a fundamental one; instead it would be derived from the standard unit of time, namely the second. According to a proposal now being considered by a committee of the International Bureau of Weights and Measures, the meter would be defined as "the distance traveled in a time interval of 1/299,792,458 of a second by plane electromagnetic waves in a vacuum". In other words, the meter would be 1/299,792,458 of a light-second.

'The meter and the second are already closely coupled. The best available method of establishing the speed of light is to compare the wavelength of electromagnetic radiation with its frequency, and thus to compare a standard of length with a standard of time. Under the proposed redefinition such experiments would still be possible, but their results would have a different interpretation. The speed of light would no longer be subject to revision; any refinement in the accuracy of measurement would alter not the velocity of light but the length of the meter.'

15 Read the following article by Philip Tajitsu Nash (from the *New York Nichibei*, 30 May 1985). Discuss the main point and the measurement concepts that the author uses to argue his viewpoint. What is your opinion, and why?

'Why not 'Oriental'?

'Has anyone ever called you an "Oriental"? Do you refer to yourself as one? For reasons of both theory and usage, this word has been in disrepute for almost two decades. It's time we all laid it to rest.

'The theoretical problem with "Oriental" stems from its Latin root words "oriri" (to rise) and "oriens" (place where the sun rises). If the earth is a sphere, there is no place where the sun rises or sets ("occidental", by the way, means place where the sun sets). Rising and setting are relative to the location of the observer. Which gets to the usage problem.

'The English language was invented by observers who came from tiny islands on the western end of the Euroasian land-mass. Our perceptions of the world are shaped by this language and by these people. For example, we are all taught to distinguish between England, Scotland, Wales and Ireland, but are not told that the Chinese cities of Shanghai, Peking, Tianjin, Canton, Shenyang, Wuhan and Chendu each have more residents than the entire 3.5 million population of Ireland. The land areas of Europe and North America, respectively, are only 7% and 16% of the earth's surface, whereas Africa is 20% and Asia is 30%. Population-wise, Asia is 60% of the world, with North America only 8%.

'The shapers of our world view and of our self-perceptions, in other words, were white European explorers, conquerors, missionaries, and others who, through historical happenstance . . . , have so far been able to impose their framework onto the rest of the world. Perfect examples are Greenwich Standard Time, named after a borough of London, and used to separate the world into 24 time zones, and the Mercator Projection, the 400-year-old concept of cramming a sphere onto a flat map that has artificially increased the size of Europe and North America in our minds (because of exaggerations at the poles).

'"Oriental" and "occidental", then, are inherently tainted by a Eurocentric viewpoint. Even if this viewpoint was non-judgmental and value-free, I could not accept its omission of Africa and South America, as well as all Native American peoples. But it is not even value-free.

'Centuries of information and misinformation about Asian peoples and countries, coupled with an exaggerated sense of self-importance for European-derived peoples, have created stereotypes about "Orientals". We are not human ("they don't value life", said a United States senator recently) and are intellectual or sexual robots. We had to be "discovered" by European explorers and "Christianized" by their missionaries. Even now our bodies don't measure up to Madison Avenue's ideals, and we must "assimilate" (stop being "Oriental") to "make it". But how can we assimilate and stop being "Oriental" when we are viewed and view ourselves as "Oriental" – something different, something from the East, something less than "white"?

'In 1968, a phrase was coined which captured the essence of the dilemma and helped to solve it while creating a new consciousness. It recognized our commonalities as Asians, distinguished us from others without the negative connotations of "Oriental", and served to reassert our ethnic pride and self-importance without denigrating anyone else's heritage. We weren't inherently better or worse, we were different. We were "Asian Americans".'

16 Discuss why the following map shows 'a new world of understanding'.

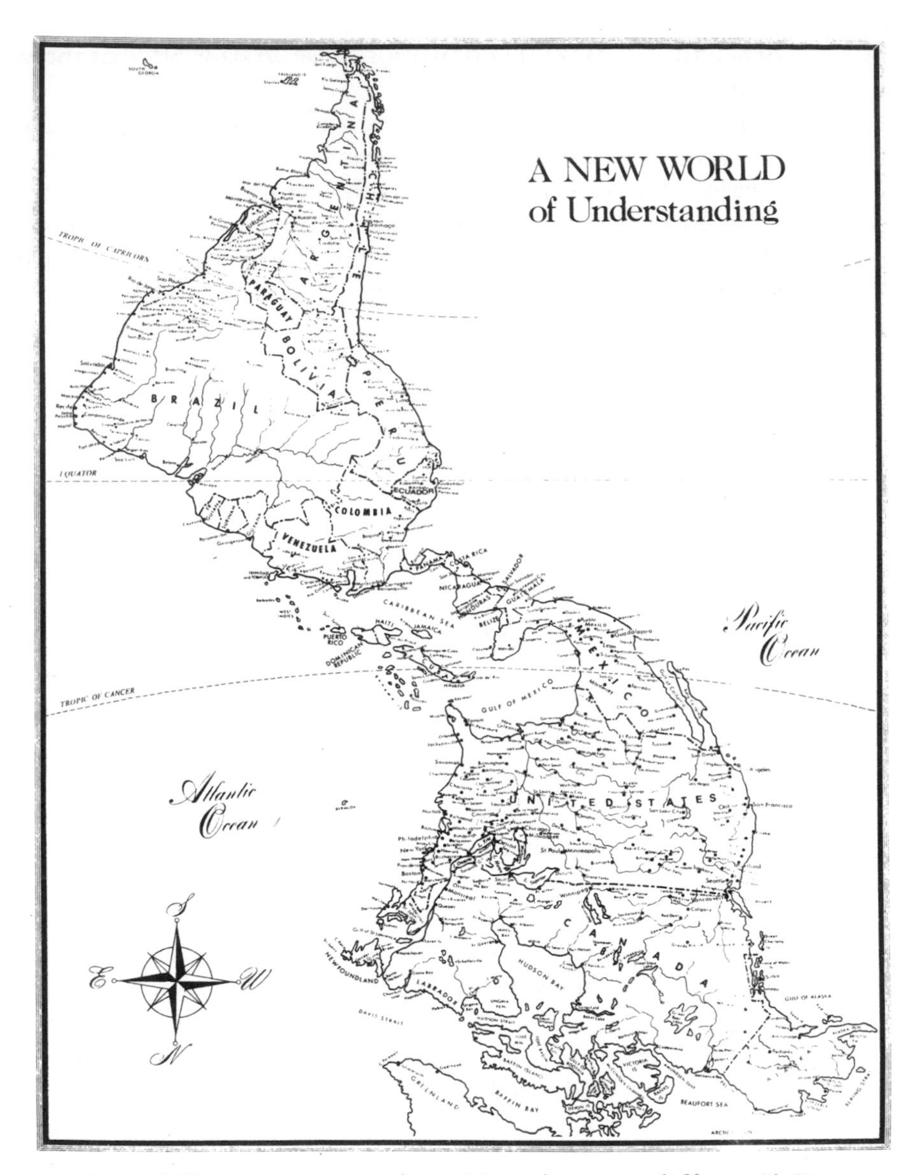

17 Read the following excerpt from 'Are the races different?' (Lewontin, 1982). Discuss how the scientific measurement data about the ABO blood group support the argument that none of the polymorphic genes perfectly discriminate one 'racial' group from another.

'It turns out that 75 per cent of the different kinds of proteins are identical in all individuals tested from whatever population, with the exception of an occasional rare mutation. These so-called *monomorphic* proteins are common to all human beings of all races, and the species is essentially uniform with respect to the genes that code them. The other 25 per cent are *polymorphic* proteins. That is, there exist two or more alternative forms of the protein, coded by alternative forms of a gene, that are reasonably common in our species. We can use these polymorphic genes to ask how much difference there is between populations, as compared with the difference between individuals within populations.

'An example of a highly polymorphic gene is the one that determines the ABO blood type. There are three alternative forms of the gene which we will symbolize by A, B, and O, and every population in the world is characterized by some particular mixture of the three. For example, Belgians have about 26 per cent A, 6 per cent B, and the remaining 68 per cent is O. Among Pygmies of the Congo, the proportions are 23 per cent A, 22 per cent B, and 55 per cent O. The frequencies can be depicted as a triangular diagram, as shown in Figure 1.

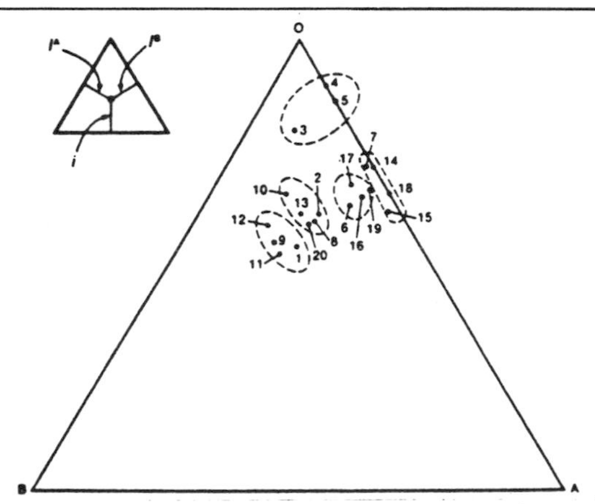

Figure 1. Triallelic diagram of the ABO blood-group allele frequencies for human populations. Each represents a population; the perpendicular distances from the point to the sides represents the allele frequencies as indicated in the small triangle. Populations 1–3 are Africans, 4–7 are American Indians, 8–13 are Asians, 14–15 are Australian aborigines, and 16–20 are Europeans. Dashed lines enclose arbitrary classes with similar gene frequencies, which do not correspond to the 'racial' classes.

Each point represents a population, and the proportion of each gene form can be read as the perpendicular distance from the point to the appropriate side of the triangle. As the figure shows, all human populations are clustered fairly close together in one part of the frequency space. For example, there are no populations with very high A and very low B and O (lower right-hand corner). The figure also shows that populations that belong to what we call major "races" in our everyday usage do not cluster together. The dashed lines have been put around populations that are similar in ABO frequencies, but these do not mark off racial groups. For example, the cluster made up of populations 2, 8, 10, 13, and 20 includes an African, three Asian, and one European population.

'A major finding from the study of such polymorphic genes is that none of these genes perfectly discriminates one "racial" group from another. That is, there is no gene known that is 100 per cent of one form in one race and 100 per cent of a different form in some other race. Reciprocally, some genes that are very variable from individual to individual show no average difference at all between major races.'

18 Study the following graphs (from the *New Internationalist*, Novem-

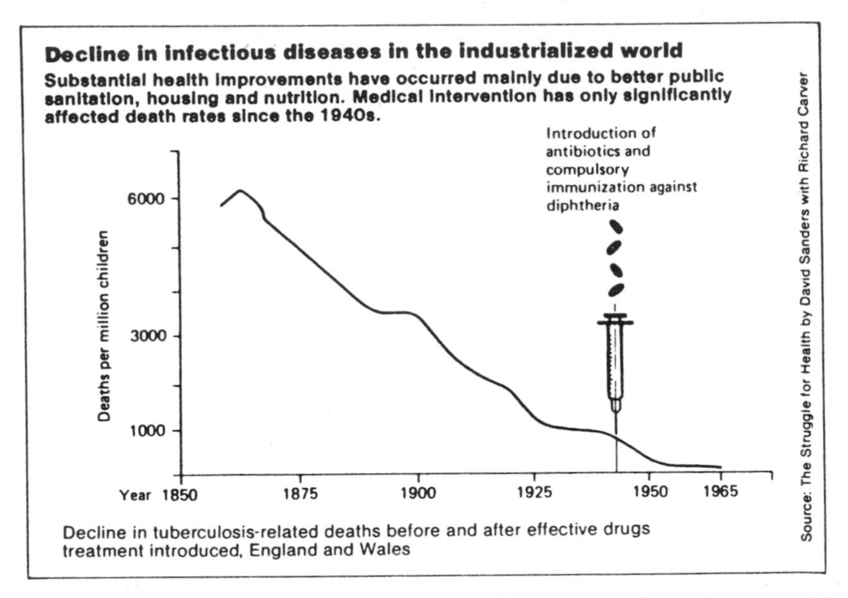

Decline in tuberculosis-related deaths before and after effective drugs treatment introduced, England and Wales

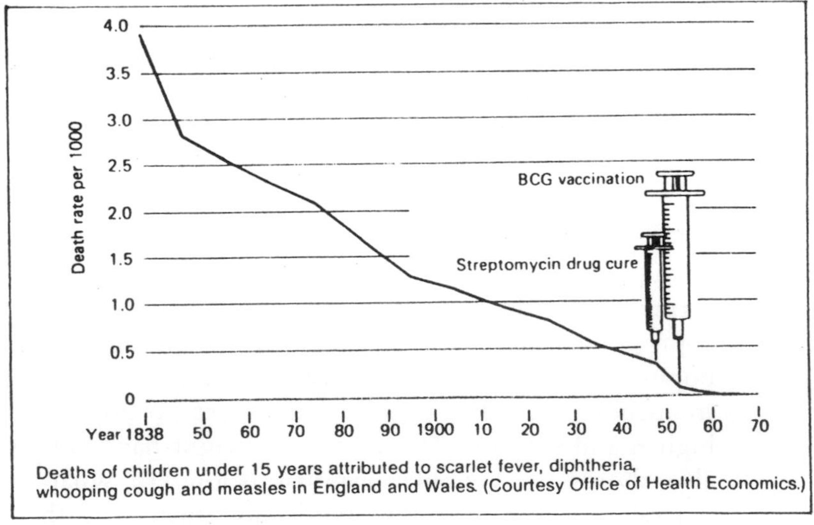

Deaths of children under 15 years attributed to scarlet fever, diphtheria, whooping cough and measles in England and Wales. (Courtesy Office of Health Economics.)

ber 1986). Discuss why the graphs provide data supporting the author's argument that 'substantial health improvements have occurred mainly due to better public sanitation, housing and nutrition'.

REVIEW QUIZ (the meaning of measurements)

1 List an item that can be measured exactly.

2 List an item that can only be measured approximately.

3 List three metric units of measurement.

4 List three 'compound' measures.

5 What is the main reason we standardize our units of measurement?

ANSWERS: CHAPTER 17

Solved problems ('you try these')

1 The amount of money you have in your pocket.

Unsolved problems

(Most of the problems do not have answers here because the possibilities are so many.)

13(a) Because the Metric Conversion Act of 1975 makes switching voluntary, the situation is confusing, since some companies are switching and others aren't. And companies that are going metric are not doing it in a way that takes advantage of the benefits of ease of calculation in the metric system.

(b) The metric system is easier to work with because the measures come in multiples of 10 so that would make it easier to compare the unit prices of variously packaged items in a supermarket.

(c) Price comparison is only easy if the actual dimensions of products are changed to metric dimensions; what's happening now is that our US measures are just being replaced by their metric equivalents. Conversion to metric of the entire US economy would cost billions of dollars. For example,

(e) $\frac{1}{2}$ gallon = 2 quarts, and 1 quart = 2 pints, and a half-gallon = 8 half-pints. In other words, each measure is a whole-number multiple of each other measure.

(f) So if the cost of the $\frac{1}{2}$ gallon is less than twice the cost of the quart, for example, the $\frac{1}{2}$ gallon is a good buy.

(g) Since (200 ml) = 3.75 times (750 ml), to find out if the 750 ml amount is a good buy you have to see if it costs less than 3.75 times the 200 ml size, so you have to multiply by a harder number than if the sizes are related by whole numbers.

(h) It's a metric version of 'an ounce of prevention is worth a pound of cure'.

14(a) The speed of light may become a defined rather than a measured quantity. Then, in order to keep the speed of light constant, 'any refinement in the accuracy of measurement would alter not the velocity of light, but the length of the meter'.

(b) 299,792,458 $\doteq$ 300 million metres per second.

(c) $\frac{1}{10{,}000{,}000}$ = 0.0000001 of the distance from the North Pole to the Equator, measured on the meridian passing through Paris.

15 The shapers of our language, and thereby of our world-view and our self-perceptions, were white Europeans who through historical situations 'have so far been able to impose their framework on to the rest of the world'. This includes measuring time from the white European vantage-point. So, for example, Greenwich Standard Time, named after a borough of London, is used to separate the world into 24 time zones. And the word 'Oriental' which from the Latin roots means 'place where the sun rises', stems from this same Eurocentrism. That is, where the sun rises is always relative to the observer's location; to fix Asia as the 'Oriental' part of the world is to take Europe as the central part, relative to which everywhere else is measured.

16 As explained in the text accompanying the map,

'Ever since maps were first drawn, certain countries have been located at the top, others below. Since "on top", "over" and "above" are equated with superiority, while "down there", "beneath" and "below" imply the reverse, these wholly arbitrary placements, over the years, have led to misconceptions and misjudgements. This Turnabout Map of the Americas serves to correct the imbalance. It focuses attention in new directions towards areas of exploding populations, energies and potentialities. It is geographically correct . . . Only the perspective has been changed.

'The map can be ordered from Laguna Sales, 7040 Via Valverde, San José, CA 95135, USA.'

17 As the scientific measurements of the proportions of the A, B and O alternative forms of blood-type gene show, 'all human populations are clustered fairly close together . . . [and] populations that belong to what we call major "races" in our everyday usage do not cluster together'. The populations most similar in ABO frequencies are not those that are normally labelled as the same racial group.

18 Medical intervention has become a significant factor only since the 1940s, yet living conditions have been improving over a much longer period of time. As the *New Internationalist* argues,

'Water-related infections were reduced by efficient sewage disposal, clean tap water and safe milk. Airborne infections like tuberculosis were reduced by better housing. And underpinning all this was a better supply of good food. This is not to deny that smallpox vaccinations or, more recently, antibiotics haven't been useful. But medicines do not lie at the basis of Western health. Decent living conditions certainly do . . .

'The fundamental causes of ill-health are beyond the control of doctors and their drugs. Yet recognizing this would mean questioning the validity of expensive medical care. It is not in the interests of the medical profession to be examining or confronting the social roots of illness. Anyway, they are not trained to be social workers or revolutionaries. They are trained to be scientists . . .

'The distortion of medical teaching is particularly acute in the Third World. Bangladeshi doctors who qualified in 1980 after five years' training in Dhaka would not have had one lecture on the appropriate non-pharmaceutical treatment for diarrhoea, though this accounted for half of all children's illness and death in the country. It is because of such inadequate training that doctors globally are prescribing over $400 million worth of anti-diarrhoeal drugs a year, the vast majority of which are useless or harmful. Doctors are reluctant to use oral rehydration salts (ORS), a cheap and simple mixture of water, sugar and salt. Every year five million children die from diarrhoea, yet 3.5 million of those deaths would have been preventable by these salts. The reason, of course, is that there is little money to be made out of ORS, but a great deal to be made out of anti-diarrhoeal drugs.'

18 The meaning of signed numbers

Although it is easy to find real-life applications of fractions, such as $\frac{1}{2}$ of an orange or $\frac{1}{4}$ of a year, applications of negative numbers are harder to grasp: what could -7 oranges or -5 years mean? Historically, as with many new 'radical' ideas, the concept of negative numbers was viewed with discomfort and suspicion for centuries. The ancient Greek mathematicians felt equations like '$6 + x = 2$' were not 'legitimate' because their solutions were negative numbers. Although negative numbers were written about as early as the Han period (200 BC to AD 200) in China, and were developed in detail by Hindu mathematicians in the seventh century, most European Renaissance mathematicians still viewed negative quantities as 'fictitious' until as late as the nineteenth century.

Morris Kline documents that most sixteenth- and seventeenth-century European mathematicians did not accept negative numbers as numbers, 'or if they did, would not accept them as roots of equations . . . Descartes . . . called negative roots of equations false, on the ground that they claim to represent numbers less than nothing . . . Pascal regarded the subtraction of 4 from 0 as utter nonsense . . . Euler, in the latter half of the eighteenth century, still believed that negative numbers were greater than infinity' (Kline, 1972, pp. 252, 593).[1]

Negative numbers gradually gained universal acceptance as the structure of mathematics became more fully developed and as their usefulness in various real-life situations became apparent. Today, we interpret -7 oranges as 'a debt of 7 oranges' and -5 years as '5 years ago'. By including the negative numbers with the positive whole numbers and positive fractions, we enlarge the types of numbers available for describing our world. The set of positive numbers, zero, and the negative numbers is called the *signed numbers*.

SOLVED PROBLEMS

Drill

In problems 1–2 replace the question mark with the comparison symbol ($<$, $>$, $=$) which will make the resulting statement true.

Problem 1 12 ? -4

Solution In order to help us visualize the relationships among the signed numbers we use a graph called a number line:

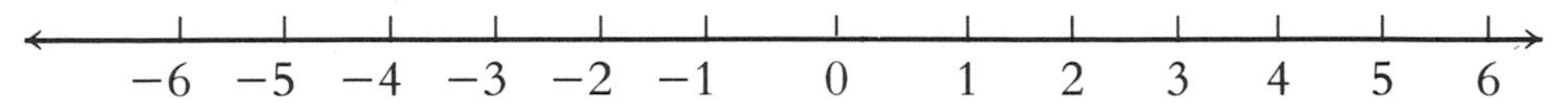

The numbers are ordered by size. The numbers get larger as we move to the right along the number line; the numbers get smaller as we move to the left along the number line. Any given number is greater than every number to its left (for example, -2 is greater than -3, greater than -4, than -5, and so on) and smaller than every number to its right (for example, -2 is smaller than -1, smaller than 0, than 3, and so on).

Also, the positive numbers are sometimes written as $+1$, $+2$, $+3$, etc. In this book any number written without a sign is understood to be positive. All positive numbers are to the right of zero; all negative numbers are to the left of zero.

The number 12 is to the right of -4 on the number line, so $12 > -4$. Also, we know $12 > -4$ because every positive number is greater than every negative number.

Answer $12 > -4$

You try this *(Answer at end of chapter)*

15 ? -3

Problem 2 -3 ? -6

Solution Again, we can use the number line to see that -3 is to the right of -6, so $-3 > -6$.

Answer $-3 > -6$

Note Do not get confused: $3 < 6$, but $-3 > -6$, the 'more negative' a number is the smaller it is. This makes sense when we consider the interpretation of negative numbers; for example, a temperature of $-6°$ is less than (or colder than) a temperature of $-3°$.

You try this *(Answer at end of chapter)*

-3 ? -15

Problem 3 Put the following numbers in size order from smallest to largest: -6, -15, 4.

On the number line -15 is furthest to the left of 0; -6 is also to the left of 0, but closer to 0 than -15. The number 4 is to the right of 0.

Answer So, the size order from smallest to largest is -15, -6, 4.

Note As discussed in Chapter 12, we can write $-15 < -6 < 4$, showing -6 is between -15 and 4. Putting the numbers in size order from largest to smallest, we can write $4 > -6 > -15$, instead of $4 > -6$ and $-6 > -15$.

You try this *(Answer at end of chapter)*

-7, -8, -6

Applications

In problems 4–5 represent each situation by the appropriate signed number.

Problem 4 Losing 5 lb.

Solution Traditionally we think of a gain as positive, and therefore we represent the opposite situation, a loss, as negative.

Answer -5 lb

Note You could argue that losing 5 lb is a 'positive' situation for an individual whose health is weak because of overweight and therefore you would represent losing 5 lb as $+5$ lb. This is not wrong – as long as you then represent the opposite situation, gaining 5 lb, as -5 lb. Positive and negative numbers are relative in the same way that left and right are relative. Something on your right may be on someone else's left, but anything opposite your right is on your left.

You try this *(Answer at end of chapter)*

Losing \$2

Problem 5 An article in the *Boston Globe* (18 January 1982) offers statistics from the second budget of Reagan's administration to show that 'the impact of President Ronald Reagan's first year is showing itself not in reduced federal spending, but in a shift of dollars away from states and cities and toward a military buildup financed largely by cuts in domestic programs'. Discretionary spending for the military was up \$30.3 billion and discretionary non-military domestic spending was cut \$25.4 billion.

Solution This is not just a textbook exercise – the article actually represented these numbers as signed numbers in the following bar graph:

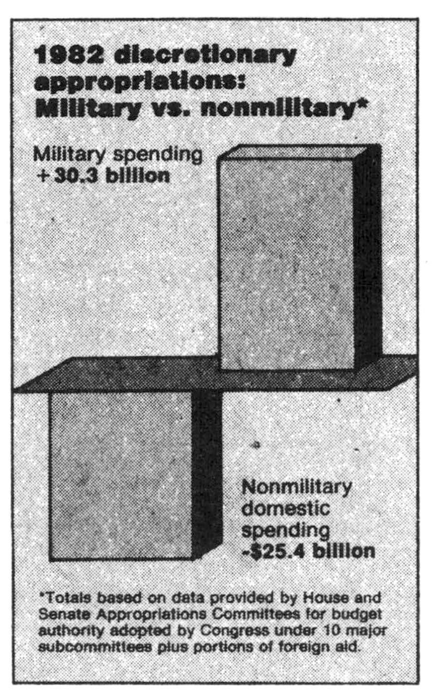

Answer Comparison of 1982 to 1981 discretionary appropriations: military spending +\$30.3 billion, non-military domestic spending −\$25.4 billion.

Note The *Globe* article indicates how the appropriations 'are an example of how the Administration used its budget cuts to change policy as much as reduce spending'. For example, while conservation and renewable energy research and development spending were cut significantly, nuclear-fission R&D spending was increased. 'As part of the nuclear emphasis, the Clinch River Breeder Reactor in Senate Majority Leader Howard Baker's home state of Tennessee received a 13% increase, but [total] solar R&D was cut by 41%.'

You try this *(Answer at end of chapter)*

Spending on child nutrition decreased by 31%.

Problem 6 According to the *1979 Information Please Almanac*, three of the lowest recorded temperatures in the world were in Antarctica at −127° F, Europe at −67° F and Alaska at −80° F. Put these temperatures in size order from coldest to warmest.

Solution -127 is furthest to the left of 0 on the number line, so it represents the coldest temperature. Since -80 is between -127 and -67 on the number line, -67 represents the warmest temperature.

Answer $-127°\text{ F} < -80°\text{ F} < -67°\text{ F}$, or $-127°$ F is colder than $-80°$ F, which in turn is colder than $-67°$ F.

You try this *(Answer at end of chapter)*

Three other lowest recorded temperatures were in Greenland at $-87°$ F, Asia at $-90°$ F, and South America at $-27°$ F. Put these temperatures in size order from warmest to coldest.

UNSOLVED PROBLEMS *(Answers at end of chapter)*

Drill

1 Replace the question mark with the comparison symbol ($<$, $>$, $=$) which will make the resulting statement true.

(a) -10 ? -3
(b) -4 ? 5
(c) 6 ? 0
(d) 0 ? -2
(e) 13 ? -1
(f) -13 ? 1
(g) 13 ? 1
(h) -13 ? -1

2 Put each group of numbers in size order from smallest to largest.
(a) $+6, -1, +1, +5$
(b) $-5, -4, -6, -10$
(c) $-5, 0, -4$
(d) $2\frac{1}{2}, 0, -\frac{1}{4}$

Thought

3 (a) Create and solve a quiz which reviews the major concepts of this chapter.
(b) Write about which of your problems are hard, which are easy, and why.

Applications

Represent each of the following situations by the appropriate signed number.

4 (a) A temperature of 30° below zero.
(b) Losing 20 lb.
(c) Sea level.
(d) 320 ft below sea level.
(e) A deficit of $142.
(f) Falling 10 ft.

5 'The Social Security Act leaves to the states the determination of the actual dollar grant paid to AFDC recipients. Each state prepares what

is supposed to be a budget of minimum needs, but these have little meaning because the state is not compelled to pay the amount budgeted and because the budgets themselves are artificially low' (Stack and Semmell, 1975, p. 91).

(a) For example, in 1972 Mississippi paid AFDC families $217 less per month than their computed per month minimum needs for those families.

(b) For another example, in 1972 Maine paid AFDC families $181 less per month than their computed needs.

6 In *Labor and Monopoly Capital*, Harry Braverman discusses the growing number of clerical workers, the decreasing amount of skill and independence involved in clerical jobs, and how this change came about as a result of monopoly capitalism (pp. 295–6). The US census of 1870 classified 0.6% of all 'gainful workers' in clerical occupations. In 1970, 18% of all workers were clerical workers – an increase of about 13,900,000 clerical workers since 1870.

The creation of a new class of workers having little continuity with the small and privileged clerical stratum of the past, is emphasized by fundamental changes in two other respects: composition by sex and relative pay.'

- In 1851 in Britain no more than 0.1% of clerks were women – by 1961, about 67% of clerks were women in Britain. From 1900 to 1970, the number of women clerical workers in the USA rose by about 9,800,000 women.
- In the USA in 1900 clerks earned $500 a year more than 'blue-collar' workers. In 1971 clerical workers earned about $20 a week less than the average 'blue-collar' worker.

7 In 'Community colleges and social stratification', Jerome Karabel presents the following table to show how 'California's three-track system of state university, state college, and community college results in poor people's paying for the education of the affluent'. Interpret the meaning of the signed numbers on the table.

Average family incomes, average higher education subsidies received, and average state and local taxes paid by families, by type of institution children attend, California, 1964

	All families	Families without children in California public higher education	Families with children in California public higher education			
			Total	Junior college	State college	University of California
1 Average family income	$8,000	$7,900	$9,560	$8,800	$10,000	$12,000
2 Average higher education subsidy per year	–	0	880	720	1,400	1,700
3 Average total state and local taxes paid	620	650	740	680	770	910
4 Net transfer (line 2–line 3)	–	−650	+140	+40	+630	+790

8 'Government fiddles while Britain burns' (*Labour Research*, May 1987) argues that: 'The British government's policy on unemployment now focuses almost entirely on reducing the official monthly figures, regardless of what happens to actual jobs and employment.' Interpret the meaning of the signed numbers in the following table from that article.

The unemployment fiddles

Date	Measure	Effect on unemployment official total
(A) Effective by end-1986		
October 1982	Basis of count changed from registered work seekers to claimants for unemployment benefit	−170,000 to −190,000 (also reduction of 100,000+ school-leavers in June to August and ending of part-time job-seeker figures −52,200)
April/June 1983	Men over sixty (a) no longer have to sign on to get insurance credits (b) automatically entitled to long-term rate of supplementary benefit	−161,400
July 1985	Northern Ireland figures reconciled with computer records	−5,000

continued

The unemployment fiddles *continued*

Date	Measure	Effect on unemployment official total
March 1986	Deduction from count of those leaving register for up to two weeks after date of count	−50,000
October 1985	Deduction of benefit for voluntary unemployed extended from six to thirteen weeks	−2,500
(B) Effective over period after late 1986		
July 1986	Restart interviews for long-term unemployed	(possible) −100,000
October 1986	Abolition of half and three-quarter rates of unemployment benefit	(possible) −24,000
October 1986	Tighter availability for work test introduced	(possible) −95,000

9 The lowest recorded temperature in Africa of −11° F occurred in 1935. The lowest recorded temperature in Australia of −8° F occurred in 1947. Which temperature is colder?

10 Some of the lowest places in the world are the Dead Sea in Asia, which is 1,290 ft below sea level; the Qattara Depression in Egypt which is 440 ft below sea level; Death Valley in California which is 282 ft below sea level; Lake Eyre in Australia which is 38 ft below sea level; and the Caspian Sea in the USSR which is 96 ft below sea level. Put the elevations of these places in size order from lowest to highest.

11 According to *The Empty Pork Barrel* (Anderson, 1979), 'There is a direct relationship between the lush budgets voted to the military and the tens of thousands of teachers who cannot find jobs. When the Pentagon's budget is about $78 billion, state and local governments have about $10 billion less to spend since 12.8% of each billion dollars going to the military would have gone towards state and local expenditures . . . On the average, teachers constitute about 23% of state and local government employees. Knowing the percentages in each state, and the amount of extra state and local government expenditures that would have taken place but for such high military budgets, it can be determined how many fewer educational jobs there are by state . . . *Every* state loses teachers' jobs.' Put the following states in size order from those losing the most teaching jobs to those losing the fewest teaching jobs.

State (partial listing)	*Number of teachers' jobs foregone*
Alabama	−22,900
California	−22,000
Maine	−1,100
Massachusetts	−7,300
Michigan	−10,200
Montana	−800
New Mexico	−1,100
North Dakota	−870
Pennsylvania	−13,100
Virginia	−4,700

12 The following chart from 'A chilling experience' (Knill and Fawcett, 1982) gives wind-chill temperatures. These are estimates of how cold a person feels given both the temperature and the wind speed. For example, if the temperature is 2° Celsius and the wind is calm, you feel as if it were 2° C (35.6° Fahrenheit). If the wind is blowing hard at 48 km/h, however, you feel as if it is −11° Celsius (+12.2° F). Put the following temperature, wind-speed pairs in size order, from coldest to warmest wind chills: −9° C and 40 km/h wind speed; −18° C and 16 km/h wind; − 29° C and 64 km/h wind; −7° C and 64 km/h wind; −4° C and 24 km/h wind; −23° C and 8 km/h wind; −34° C and calm wind.

Wind-chill chart

Wind speed	Thermometer reading (degrees Celsius)														
	4	2	−1	−4	−7	−9	−12	−15	−18	−21	−23	−26	−29	−32	−34
Calm	4	2	−1	−4	−7	−9	−12	−15	−18	−21	−23	−26	−29	−32	−34
8 km/h	3	1	−3	−6	−9	−11	−14	−17	−21	−24	−26	−29	−32	−36	−37
16 km/h	−2	−6	−9	−13	−17	−19	−23	−26	−30	−33	−36	−39	−43	−47	−50
	Cold				**Very**										
24 km/h	−6	−9	−12	−17	−21	−24	−28	−32	−36	−40	−43	−46	−51	−54	−57
					cold		**Bitter**					**Extremely**			
32 km/h	−8	−11	−16	−20	−23	−27	−31	−36	−40	−43	−47	−51	−56	−60	−63
							cold					**cold**			
40 km/h	−9	−14	−18	−22	−26	−30	−34	−38	−43	−47	−50	−55	−59	−64	−67
48 km/h	−11	−15	−19	−24	−28	−32	−36	−41	−45	−49	−53	−57	−61	−66	−70
56 km/h	−12	−16	−20	−25	−29	−33	−37	−42	−47	−51	−55	−58	−64	−68	−72
64 km/h	−13	−17	−21	−26	−30	−34	−38	−43	−48	−52	−56	−60	−66	−70	−74

REVIEW QUIZ (the meaning of signed numbers) *(Answers at end of chapter)*

1 Replace the question mark with the comparison symbol (<, >, =) which will make the resulting statement true: $-12 ? -12\frac{1}{2}$.

2 Put the following numbers in size order from smallest to largest:

$-\frac{1}{2}, -\frac{1}{3}, \frac{1}{2}, 0, \frac{1}{3}$

3 Represent each of these situations by the appropriate signed number:
(a) Climbing a 4200-ft mountain
(b) Paying a bill for $28.75.

4 Interpret the meaning of the signed numbers in the following table (from the *Boston Globe*, 18 January 1982, as described in 'Solved Problems: Applications', problem 5, earlier in this chapter).

Program	1981	1982	Difference
Non-Military Expenditure			
Title XX	$2.99 billion	$2.4 billion	(−19%)
Community health	1.02 billion	0.788 billion	−22%
Community services administration	0.525 billion	0.348 billion	−33%
Preventive health	0.338 billion	0.284 billion	−16%
Health planning	0.126 billion	0.062 billion	−50%
Head start	0.818 billion	0.912 billion	+11%
Military expenditure			
Procurement	$48.0 billion	$64.8 billion	+35%
Operation and maintenance	55.9 billion	61.85 billion	+10%
Personnel	36.87 billion	38.1 billion	+3%
Military construction	5.3 billion	7.06 billion	+32%
Atomic energy	3.67 billion	4.67 billion	+28%
Foreign military aid	0.638 billion	0.965 billion	+51%

ANSWERS: CHAPTER 18

Solved problems ('You try these')

1 $15 > -3$
2 $-3 > -15$
3 $-8 < -7 < -6$
4 $-\$2$
5 -31%
6 $-27°$ is warmer than $-87°$ which is warmer than $-90°$, or $-27 > -87 > -90$.

Unsolved problems

1(a) $-10 < -3$
(b) $-4 < 5$
(c) $6 > 0$
(d) $0 > -2$
(e) $13 > -1$
(f) $-13 < 1$
(g) $13 > 1$
(h) $-13 < -1$
2(a) $-1 < +1 < +5 < +6$
(b) $-10 < -6 < -5 < -4$
(c) $-5 < -4 < 0$
(d) $-\frac{1}{4} < 0 < 2\frac{1}{2}$
4(a) $-30°$
(b) -20 lb
(c) 0
(d) -320 ft
(e) $-\$142$
(f) -10 ft
5 Value of 1972 AFDC grants when compared to computed monthly minimum needs:

(a) Mississippi	−$217
(b) Maine	−$181

'Families that try to raise their total income through earnings face the frustration of seeing most of their earnings go to the state and federal governments in the form of reduced AFDC payments.'

6 In the USA from 1870 to 1970, the change in the number of clerical workers is +13,900,000.

In the USA from 1900 to 1970, the change in the number of women clerical workers is +9,800,000.

US comparison of clerical workers to blue-collar salaries:

1900	+$500 per year
1971	− $20 per week

The giant increase in clerical labour has come about because of the complexities of keeping track of value in monopoly capitalism. 'From the point of view of capital, the representation of value is more important than the physical form or useful properties of the labor product. The particular kind of commodity being sold means little; the net gain is everything. A portion of the labor of society must therefore be devoted to the accounting of value.' Braverman goes on to show that the labour expended transforming the commodity into money value, including cashiers and collection work, the record-keeping, accounting, and so on, 'begins to approach or surpass the labor used in producing the underlying commodity or service . . . entire "industries" come into existence whose activity is concerned with nothing but the transfer of values . . . [and] since the work of recording the movement of values is generally accomplished by a capitalist agency for its particular ends, its own accounting has no standing with other organizations. This leads to an immense amount of duplication . . . Thus the value-form of commodities separates itself out from the physical form as a vast paper empire which under capitalism becomes as real as the physical world, and which swallows ever-increasing amounts of labor . . . certainly there is no doubt that the demands of marketing, together with the demands of value accounting, consume the bulk of clerical time.'

7 Families with no children in California public higher education lose $650 per year since they pay that amount in taxes. Junior-college families whose average income is $8,800 gain an average of $40 per year since they pay $680 in taxes and get an average subsidy of $720. As average income increases, children go to the more prestigious institutions and families' subsidies increase – for example, university families gain $790 per year. 'The higher ranking the institution, the more public money spent on the students . . . [and] eligibility to attend the high-cost state university or middle-cost state college system shows a generally linear relationship with family income . . . Since taxes in California are approximately proportional to income and since the probability of attending an expensive school is positively related to income, the net effect of the California system is to redistribute income from poor to rich.'

8 The table shows the reductions in the official British unemployment rate created by changes in the way the unemployed are counted. For example, by not counting those leaving the unemployment register for up to two weeks after the date of the count, the unemployed count is reduced by 50,000 people. The article reports that none of the changes effective by the end of 1986, however, reflected any real change in the numbers actually out of work.

9 $-11°$ F is colder.

10 $-1{,}290 < -440 < -282 < -96 < -38$

11 Alabama loses the most teaching jobs, then California, Pennsylvania, Michigan, Massachusets, Virginia, Maine and New Mexico lose the same amount, then North Dakota, then Montana loses the fewest (800 jobs). 'In the fall of 1976, about 186,000 recent college graduates, their newly won teaching certificates in their hands, went out to look for jobs. 92,000 of them could not find a teaching job. The amount of money it would cost to put these 92,000 young teachers to work is spent by the Pentagon between Monday morning and Thursday evening of any given week.'

12

	°C	wind speed (km/h)	wind chill
Coldest:	−29	64	−66
	−34	calm	−34
	−18	16	−30
	− 9	40	−30
	− 7	64	−30
	−23	8	−26
Warmest:	− 4	24	−17

Review quiz

1 $-12 > -12\frac{1}{2}$

2 $-\frac{1}{2} < -\frac{1}{3} < 0 < \frac{1}{3} < \frac{1}{2}$

3(a) +4,200 ft

(b) −$28.75

4 Community health spending decreased 22%; Community Services Administration

has been cut 33%; preventive health cut 16%; health planning cut in half; only Head Start increased 11%. The military, on the other hand, increased in all categories (procurement 35%, operation and maintenance 10%, personnel 3%, construction 32%, atomic energy 28%, and foreign military aid 51%).

19 The meaning of variables

Algebra, using unknown quantities, or variables, is a generalization of arithmetic and therefore increases the techniques you have for solving problems. Although most of the real-life problems in this text can be solved with arithmetic, many of them can be solved much more clearly and efficiently with some understanding of the use of variables in algebra. Also, because algebra is an abstraction of arithmetic, thinking algebraically gives you important experiences in abstract thinking, and abstract thinking is vital in order to develop your numerical analysis of specific situations into a general, total picture of the way society is structured. In order to see through the myths with which we were all raised and to see the real possibilities for change, you need a lot of experience thinking theoretically.

The first clear textbook on algebra, written by the Arab scholar Al-Khwarizmi in 825, was called *Al-jabr w'al-muqabalah* or 'the art of bringing together unknowns to match a known quantity'. According to Arndt (1983), '. . . in looking at mathematical texts over the years . . . we can see Al-Khwarizmi's profound effect on mathematics . . . Over the centuries, writers on elementary algebra have borrowed his format, his terminology, and his classification of the types of linear and quadratic equations; often they have even used the very same numerical examples he did . . .'[1] Interestingly, in medieval times, 'algebraist' referred either to a surgeon (who brought bones together) or to a specialist in equations (who brought unknowns and numbers together).

Equally interesting is the fact that some of the great mathematicians who contributed to the development of algebra were women. Hypatia, a Greek mathematician of the fourth century, was a renowned mathematician, philosopher, and teacher. According to a historian of mathematicians, 'she is entitled to a conspicuous place among such luminaries of science as Ptolemy,

Euclid, Apollonius, Diophantus and Hipparchus' (Mozans, quoted in Osen, 1974, p. 32). She was barbarously murdered by a mob of Christian fanatics, torn from her chariot, 'stripped naked, slashed to death with oyster shells, and finally burned piecemeal. In the eloquent, beautiful, and brilliant high priestess of mathematics and philosophy, the bigoted could see nothing but a pagan influence . . .' (Kramer, 1970, p. 64). Another woman algebraist, Emmy Noether (1882–1935), a Jew who fled to the United States to escape Hitler's persecution, was judged by Albert Einstein as 'the most significant creative mathematical genius thus far produced since the higher education of women began'. He goes on to say that, 'in the realm of algebra in which the most gifted mathematicians have been busy for centuries, she discovered methods which have proved of enormous importance . . .' (Osen, 1974, p. 151).

Modern algebraic symbolism was not widely used until the sixteenth century. '. . . for many centuries mathematicians dealt with problems involving numbers on entirely individualistic lines. Each writer would use a shorthand which he [sic] understood himself without attempting to introduce a universal convention' (Hogben, 1965, p. 306). Problems were solved by algebraic concepts, but the steps of the solutions were carried out verbally or written out in full, without the algebraic symbolism we use today. As algebraic symbolism was perfected, it became much clearer how to use algebraic concepts to solve problems. Interestingly, one of the first people to use algebraic symbols for solving equations was a merchant's clerk, not an élite academic.[2] 'What is called solving an equation is putting it in a form in which its meaning is obvious. The rules of algebra tell us how to do this. The really difficult step consists in translating our problem from the language of everyday life to the language of algebra' (Hogben, p. 308).

The following table reviews some common English phrases for the arithmetic operations which you will study in Volume 2:

<table>
<tr><td>'The sum of five and six'
'Five increased by six'
'Six more than five'
'Six added to five'</td><td>all translate '5 + 6'</td></tr>
<tr><td>'The difference of ten and four'
'Ten decreased by four'
'Four subtracted from ten'
'Four taken away from ten'
'Four less than ten'</td><td>all translate '10 − 4'</td></tr>
<tr><td>'The product of eight and five'
'Eight multiplied by five'
'Eight times five'</td><td>all translate '8 × 5'</td></tr>
</table>

'The quotient of fourteen and seven'
'Fourteen divided by seven'
'Seven divided into fourteen'
} all translate '$7\overline{)14}$' or '$14 \div 7$' or '$\frac{14}{7}$'

In addition to the translation of the arithmetic operations, two algebraic concepts are necessary for solving problems using algebra: the equation and the variable.

The equation '$7 + 5 = 12$' translates to
{ 7 + 5 equals 12
7 + 5 is 12
7 + 5 is the same as 12
7 + 5 is another name for 12

Any mathematical sentence with an equal sign is called an *equation*. An equation consists of an expression on the *left side* equal to another expression on the *right side*. Here '7 + 5' is the left side of the equation and '12' is the right side of the equation. You can visualize an equation as a balanced scale, with the left side representing weights in the left balance pan and the right side representing weights in the right balance pan.

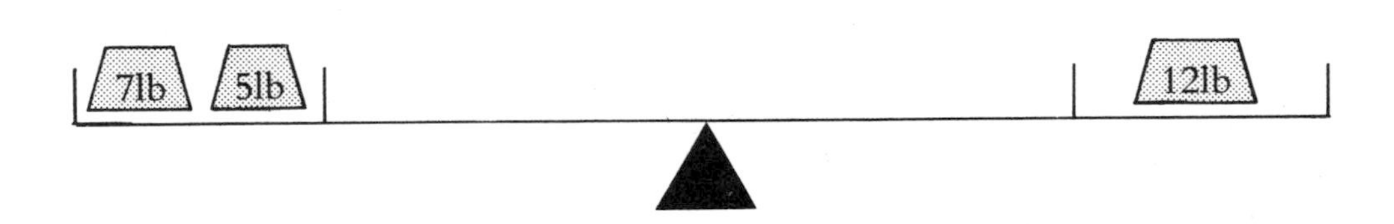

In this case, the equation $7 + 5 = 12$ tells us that a 7-lb weight and a 5-lb weight exactly balance a 12-lb weight.

A *variable* is any symbol, usually a letter like x, y, a, or n, that is used to represent one or more numbers. This section explores the use of variables in equations: to express a general mathematical rule or pattern; to show a constant relationship between several quantities (called a formula); or, to represent the unknown quantity we are looking for to solve a problem.

SOLVED PROBLEMS

Drill

Problem 1 Write the following English sentence as an equation:

5 plus 2 is the same as 7.

Solution 'plus' translates as '+'

'is the same as' translates as '='

Answer $5 + 2 = 7$

Note We can check the truth of this equation by adding 5 and 2, which sums to 7. To translate, however, just means to write the sentence with algebraic symbols instead of with words – you don't have to check the truth, and it doesn't even have to be true. For example, '7 plus 9 equals 6' translates '$7 + 9 = 6$', clearly a false equation, but a correct translation.

You try this *(Answer at end of chapter)*

3 minus 3 is another name for zero

Problem 2 Translate into English:

$$4(2) = 8$$

Solution There are many different ways of indicating the operation of multiplication:

$$\left.\begin{array}{l} 4 \times 2 = 8 \\ 4 \cdot 2 = 8 \\ 4(2) = 8 \end{array}\right\} \text{ all translate 4 multiplied by 2 equals 8}$$

In algebra, the '×' is not used because it would be confused with the variable x; the dot is not preferred because it can be confused with a decimal point; parentheses are most commonly used in algebra to indicate multiplication.

Answer 4 times 2 equals 8, or 4 multiplied by 2 equals 8

You try this *(Answer at end of chapter)*

$$3(7) = 21$$

Problem 3 Translate into English:

$$\tfrac{10}{2} = 5$$

Solution In an equation, there are two possible ways of representing division:

$$\left.\begin{array}{l} 10 \div 2 = 5 \\ \tfrac{10}{2} = 5 \end{array}\right\} \text{ translate 10 divided by 2 equals 5}$$

In algebra, we almost always represent division the second way, as a fraction.

Answer 10 divided by 2 equals 5

Note We could also read this as a fraction: '10 halves equals 5 wholes'

You try this *(Answer at end of chapter)*

$\frac{15}{15} = 1$

Problem 4 My colleague Frank Davis uses the following arithmetic schema to introduce students to translating numerical information involving variables into equations. Write the following series of arithmetic operations as an equation:

Solution These two arithmetic problems and the arrow are really a set of instructions: find the sum of 10 and 13, and then multiply that sum by 2 to get your answer, x.

> The 'sum of 10 and 13' translates '10 + 13'.
> '2 multiplied by that sum' translates '2(10 + 13)'

where the parentheses indicate multiplication *and* that you add 10 + 13 first, before multiplying the sum by 2. '$2 \cdot 10 + 13$' would mean that you multiplied 10 by 2 and then added 13. So $2(10 + 13) = 2(23) = 46$, whereas $2 \cdot 10 + 13 = 20 + 13 = 33$.

Answer Frank's instructions, therefore, tell us that $x = 2(10 + 13)$

Notes

1. When you translate, you are, in a sense, working backwards from the variable. First you see x is obtained by multiplying 'something' by 2. Then you see that that 'something' is obtained by adding 10 and 13.
2. $x = (10 + 13)\,2$ is also a correct translation since you can multiply in any order. But, as indicated above, $x = 2 \cdot 10 + 13$ is incorrect, because Frank's instruction code tells us to multiply the *sum* of 10 + 13 by 2.
3. Finding x in equations of this type just requires using arithmetic. $x = 2(10 + 13)$ means $x = 2(23)$ which means $x = 46$.

You try this *(Answer at end of chapter)*

$$\begin{array}{r} 14 \\ -\ 8 \\ \hline \Delta \end{array} \longrightarrow \quad \begin{array}{r} y \\ \Delta\,\overline{)12} \end{array}$$

Problem 5 Using any letter as the variable, write the following equation explaining what the variable represents:

6 less than what number equals 10?

Solution We can pick any letter to represent the unknown number. Let m = the unknown number. As we reviewed at the beginning of this chapter, '6 less than m' translates 'm − 6', so

Answer $m - 6 = 10$

Notes

1. You can use any letter as the variable so $y - 6 = 10$ or $n - 6 = 10$, etc. are also solutions to this problem.
2. To find the variable m in this equation involves more than just arithmetic. Because the numbers are small whole numbers you can probably see that the unknown number $m = 16$ (because 6 less than 16 equals 10). However, many equations, where the variable is not a whole number and is 'in the midst' of numbers, cannot be easily solved mentally. If you study algebra after this text, you'll learn procedures for solving many types of equations.

You try this *(Answer at end of chapter)*

5 more than 8 equals what number?

Problem 6 Translate into English:

$4z = 20$

Solution In algebra we represent multiplication of a variable and a number by just writing them next to each other without any symbol.

'$4z$' translates as '4 times z' or '4 multiplied by z'

Answer 4 times a number equals 20.

Notes

1. Since an equation with one variable often arises from a problem in which we want to find a missing number, we can translate the equation as a question: 4 multiplied by what number is 20?
2. Since we can multiply numbers in any order we could write $4z$ or $z4$. But we agree in algebra always to write the number first and the variable second, as in $4z$.
3. As above, you can probably see the answer. Since 4 times 5 equals 20, z must equal 5. Although many very basic equations can be solved mentally, you need rules for the more complicated equations.
4. In algebra we represent an unknown, x, divided by a given number, say 3, as $\frac{x}{3}$ or $\frac{1}{3x}$ (since, as we'll see in Volume 2, dividing by 3 is the same as multiplying by $\frac{1}{3}$).

You try this *(Answer at end of chapter)*

$3y = 27$

Applications

Problem 7 Use variables to write an equation which generalizes the arithmetic rule illustrated by the following specific cases:

$$23 + 0 = 23$$
$$2 + 0 = 2$$
$$1{,}256 + 0 = 1{,}256$$

Solution Each equation tells us that if we add 0 to a number, our answer is that same number. Algebra gives us a way of stating the general rule: adding 0 to any number results in that same number.

Answer Let x = any number. Then, the general rule is that $x + 0 = x$.

Notes

1 We can think of the variable as a 'box' to hold the place of a number – for any number in the box: $\triangle + 0 = \triangle$. You can visualize actually putting specific numbers in the box. For example,

$$\triangle_{5} + 0 = \triangle_{5}$$

$$\triangle_{28} + 0 = \triangle_{28}$$

Once you have chosen a number to put in the box, you must stick with that number throughout the problem. For example, once we place 5 in the $\triangle$ on the left side, we must use 5 in the box on the right side of the equation. In the next problem, of course, the box can hold a different number, such as 28.

2 One of the reasons algebra was developed was in response to the practical need for rules to make calculations quickly and easily (Hogben, 1965, p. 303). For example, algebra allows us to state the rule for multiplying fractions very succinctly:

$$\frac{a}{b} \cdot \frac{c}{d} = \frac{ac}{bd}$$

You try these *(Answers at end of chapter)*

$$12 - 0 = 12$$
$$3 - 0 = 3$$
$$172 - 0 = 172$$

Problem 8 Explain in words what relationships the following formula represents:

$d = rt$, where
d = distance travelled
r = rate, or speed at which you travelled
t = time for which you travelled

Solution In algebra, multiplication is indicated when variables are written next to each other with no operation symbol in between.

Answer The distance you travel is equal to your speed multiplied by the time you were travelling.

Notes

1 A specific example of this rule is that if you drive at 50 mph for 3 hours, the distance you've travelled is $50 \times 3 = 150$ miles.
2 Volume 2 explores the uses of formulas in greater detail.

You try this (*Answer at end of chapter*)
$A = \frac{1}{2}bh$, where A = area of a triangle, b = the length of the triangle base, h = the length of the triangle height

Problem 9 According to Mandell (1975), in 1969 only $\frac{9}{100}$ of the money collected by the Boston United Fund went to agencies serving the inner city. If n dollars were collected by the Boston United Fund in 1969, represent the amount of money that went to agencies serving the inner city.

Solution '$\frac{9}{100}$ of the money collected' translates to $\frac{9}{100}n$'

Answer $\frac{9}{100}n$ dollars went to agencies serving the inner city.

Notes

1 The 'of' translates as multiplication.
2 $\frac{9}{100}n$ can be written as $\frac{9n}{100}$.
3 Because of this unequal distribution of funds, Blacks in Boston challenged the United Fund and because 'United Fund authorities were unresponsive to this challenge, eight Black community groups initiated their own United Black Appeal with a goal of $4 million' (Mandell, p. 77).
4 Mandell further discusses the idea that 'Philanthropy may be the very glue that keeps capitalism from falling apart at the seams . . . if there were no charity to mask the gross inequalities of income and wealth, people might catch on to what is happening and get angry enough to change it' (p. 66). She documents that 'much of private

philanthropy is dominated by corporate élites, and that this generally sets conservative limits to the kinds of services that institutions give' (p. 67). The corporations don't just gain goodwill and quiet among the working class from their philanthropy. Mandell quotes Alison Arnold in the *Boston Globe* (25 January 1974) to show how important charity is to corporations: 'Judging by the number of benefit events, you might think the capitalist system has broken down completely. It looks as if the rich, in Robin Hood fashion, are robbing themselves to help the poor. But, it really doesn't work that way. As one person said: "I dread to think what would happen to us if the income tax laws were amended"' (p. 70).

You try this *(Answer at end of chapter)*

The article points out that even the $\frac{9}{100}$ portion is misleading since a good part of that money went to area-wide groups such as the Salvation Army. Only about $\frac{4}{100}$ of the money collected by the Boston United Fund went to indigenous organizations in Boston's non-white community. Represent the amount of money that went to these indigenous organizations.

Problem 10 Describe a situation that could be represented in algebra by $3y$.

Solution '$3y$' translates as '3 times y' or 'a number multiplied by 3'.

Answer If a shirt cost y dollars, then 3 of the shirts would cost $3y$ dollars.

Notes

1 This problem may be hard to do before studying the meaning of the operations in Volume 2. If so, try to follow the solution for now and consider this a preview of what will be studied in more detail in the next volume of this text.

2 Of course, there are many other situations that could be represented by $3y$. For example, if a shirt cost y dollars and another shirt cost 3 times as much, then the more expensive shirt cost $3y$ dollars. For another example, if a shirt cost \$3 and you buy y shirts, you'll be spending $3y$ dollars.

You try this *(Answer at end of chapter)*

Describe a situation that could be represented in algebra by $15 - n$.

Recreation
Problem 11 (a) Write two more lines in the following number pattern:

$1\frac{1}{3} \times 4 = 1\frac{1}{3} + 4$

$1\frac{1}{4} \times 5 = 1\frac{1}{4} + 5$

$1\frac{1}{7} \times 8 = 1\frac{1}{7} + 8$

(b) Use variables to write the general equation representing this pattern.

Solution (a) First, you might want to check that the given equations are true – they are! Then, to analyse a general number pattern from some instances (in this case three lines), examine the similarities and the differences in each instance. In this case, two numbers multiplied are equal to the same two numbers added. The first is a mixed number whose whole-number part is always 1 and whose numerator is always 1. Finally, the second number is always 1 more than the denominator.

Answer (a) $1\frac{1}{9} \times 10 = 1\frac{1}{9} + 10$

$1\frac{1}{15} \times 16 = 1\frac{1}{15} + 16$

Solution (b) To write the general equation, we must first determine what part of the pattern is variable. In this case, the denominator varies, so let n = the denominator. Then the second number, which is always 1 more than the denominator, can be represented by $n + 1$.

Answer (b) $1\frac{1}{n} \times (n + 1) = 1\frac{1}{n} + (n + 1)$

Notes

1 The parentheses are used for clarity – to show that we are multiplying and adding two numbers,

$1\frac{1}{n}$ and $(n + 1)$.

2 The pattern is quite interesting, since it's not generally true that the product of two numbers is equal to their sum!

You try this *(Answers at end of chapter)*

$2 \times 4 = 3 \times 3 - 1$
$3 \times 5 = 4 \times 4 - 1$
$4 \times 6 = 5 \times 5 - 1$

UNSOLVED PROBLEMS

(*Answers at end of chapter*)

Drill

1 Write each of the following as a numerical equation:
 (a) 6 divided by 18 equals $\frac{1}{3}$.
 (b) 8 minus $\frac{1}{2}$ is another name for $3\frac{1}{2}$ plus 4.
 (c) Six multiplied by four is equal to 48 divided by 2.
 (d) $5\overline{)126.0}$ with quotient 25.2

2 Translate each of the following numerical equations into an English sentence.
 (a) $3(15) = 45$
 (b) $\frac{123}{123} = 1$
 (c) $4(3)(2) = 3(2)(4)$
 (d) $\frac{30}{30} = 0.5$

3 Write each of the following arithmetic schemas as an equation:

(a)
 13
 +20
 ———
 △

 $11\overline{)\triangle}$ with quotient m

(b)
 $4\overline{)12}$ with quotient □

 □
 ·6
 ———
 w

(c)

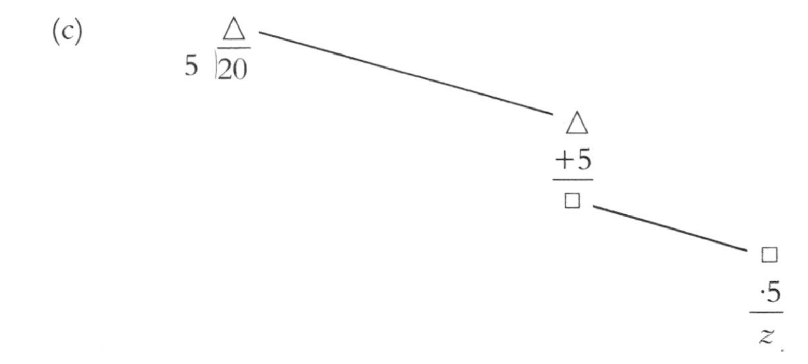

(d)

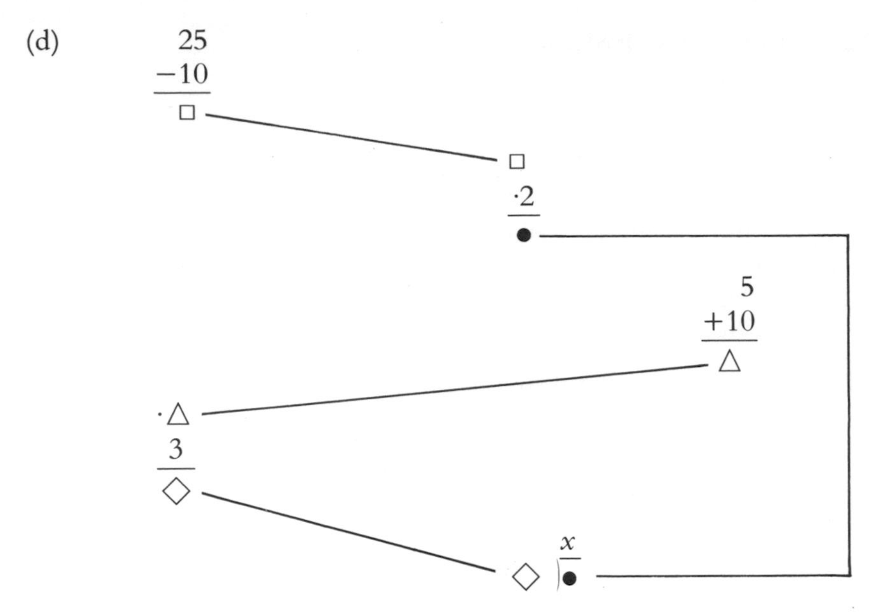

4 Write each of the following as algebra equations, using any letter as the variable, explaining what the variable represents:
   (a) Twelve plus a number is 19.
   (b) Six less than 13 is what number?
   (c) What number is 90 divided by 2?
   (d) 5 multiplied by what number is equal to 95?

5 Translate each of the following algebra equations into English:
   (a) $3y = 15$
   (b) $\frac{x}{2} = 16$
   (c) $\frac{1}{2x} = 16$
   (d) $y + 6 = 2y$

Thought

6(a) Write briefly about what was the most and what was the least interesting thing you learned in this chapter and why.
   (b) Write briefly about which was the hardest, which the easiest solved problem and why.

7(a) Create and solve your own review quiz of this chapter. You should have at least five problems, and your problems should reflect the various topics covered in this chapter.
   (b) For each problem, discuss why you chose to create it (which concept does it review, for example), and evaluate whether the problem is hard or easy and why.

Applications

8 Write an equation which generalizes the arithmetic rule illustrated by

the following specific cases:

(a) $3(1) = 3$
$1.25(1) = 1.25$
$\frac{3}{5}(1) = \frac{3}{5}$
(b) $\frac{15}{15} = 1$
$\frac{4}{4} = 1$
$\frac{1,234}{1,234} = 1$
(c) $3(0) = 0$
$15.68(0) = 0$
$4\frac{2}{3}(0) = 0$
(d) $\frac{2}{3} \times \frac{3}{2} = 1$
$\frac{4}{9} \times \frac{9}{4} = 1$
$\frac{8}{15} \times \frac{15}{8} = 1$

9 Write the rule for dividing two fractions,

$\frac{a}{b}$ and $\frac{c}{d}$ as an algebra equation.

10 As a difficult challenge, write the rule for adding two fractions as an algebra equation.

11 Explain in English what relationships each of the following formulas represent.

(a) $A = lw$, where A = the area of a rectangle
l = the length of the rectangle
w = the width of the rectangle

(b) $d = \frac{c}{\pi}$, where d = the diameter of a given circle
c = the circumference of that given circle
$\pi \doteq 3.14$

Represent in algebraic language, using the variable in problems 12–17.

12 According to *In These Times* (14–20 October 1981), 'Pesticide use has increased tenfold in the last thirty years – yet twice as many crops have been lost to pests.' If *n* tons of pesticide were used thirty years ago, represent the amount of pesticide in use as of the date of this report.

13 A survey of foundations in the USA (Mandell, p. 81) asked what percentage of their grants could be considered controversial by some, between 1966 and 1968. Foundation executives responded as follows:

	%
Voter registration	0.1
Studies of subjects directly related to public issues, dissemination of information	0.3
Community or neighbourhood organizing of an ethnic ghetto, or impoverished group	1.5
Birth control	0.9
Sex education	less than 0.05
Urban youth groups (including gangs)	1.3
Student organizations	0.8

Only 1%, or $\frac{1}{100}$, of the foundations viewed any of their grants as controversial, and the amount involved was only 0.1% or $\frac{12}{1,000}$ of the total grants made. If there were a total of *g* grants made, represent the amount of controversial grants made.

14 A *Boston Globe* article (16 December 1981) reported that the Public Service Electric & Gas Co. 'will not complete a nuclear power plant because of reduced energy demand and problems in getting favorable financing'. They asked the Board of Public Utilities for permission to charge their customers 'an extra $370 million over 12 years to cover the cost of already completed construction and the cost of abandoning the Hope Creek II plant'. If they have *x* customers, represent the average amount of welfare each customer will pay to PSE&G for their business mistake.

15 Eugene Debs wrote in 1904, three years after the formation of the Socialist Party, that 'the members of a trades union should be taught . . . that the labor movement means . . . infinitely more than a paltry increase in wages . . . that while it engages to do all that possibly can be done to better the working conditions of its members, its higher object is to overthrow the capitalist system of private ownership of the tools of labor, abolish wage-slavery and achieve the freedom of the whole working class and, in fact, of all mankind . . .' (Zinn, 1980, p. 332). Zinn goes on to give facts and figures that demolish the myth that the Socialist Party did not have wide appeal. At one time the party had 100,000 members and its main newspaper had half a million subscribers. 'The strongest Socialist state organization was in Oklahoma, which in 1914 had 12,000 dues-paying members . . . and elected over a hundred socialists to local office, including six to the Oklahoma state legislature.' At one time the party had 1,200 office holders throughout the country (p. 332). If they held office in *t* municipalities, represent the average number of office holders in each municipality.

16 In 1978 (Edelman, p. 57), the median family income of black children was about half of the median family income of white children. If the median family income of black children was x dollars, represent the median family income of white children.

17 A *Boston Globe* article (15 December 1981) reported that many corporations paid little or no US tax on their US income in 1980. '. . . based on reports by the companies themselves to the Securities and Exchange Commission . . . Chase Manhattan Bank, Squibb Corporation, and Monsanto Company paid no taxes on US income despite domestic incomes of $207 million, $78 million, and $204 million respectively.' The nation's largest company, Exxon, paid a US rate of 1.3% on their domestic incomes. If they had an income of m dollars and paid z dollars in taxes, represent their after-tax US income.

18(a) Represent the number of cents in d dimes. (1 dime = 10 cents)
(b) Represent the number of days in r weeks.
(c) Represent the number of hours in y days.

19 What's so funny about this cartoon?
Translate the Pentagon order into an equation.

20 Use any of the facts in the following article (*In These Times*, 14 November 1981) to create and solve a problem involving representing a relationship in algebraic language.

'What's in a number?

'Some day, after each of us has been stamped with a marketing code, strategically placed electronic eyes will take the uncertainty out of crowd estimates. But until then the figures for events such as the Sept. 19 "Solidarity Day" demonstration will continue to be disputed. The AFL-CIO, which sponsored the anti-Reagan rally, claims that there were more than 400,000 participants; the US Park Service and its Park police, under the jurisdiction of (yes, again) Interior Secretary James Watt, fed the major media the generally accepted estimate of 260,000.

'"The conviction runs high around here that the Park police lowballed us," a federation spokesman told "In Short" stringer Calvin Zon. "They weren't about to have it said that the largest political demonstration in US history was aimed against Ronald Reagan." The spokesman said that shortly after the speeches began around 2 p.m., with the rally still growing, a Park Service officer stationed at the speakers' platform estimated the crowd at 315,000. But throughout the afternoon, reporters calling Park Service headquarters were given the 260,000 figure.

'The Park Service denies any meddling with the numbers, but Sam Jordan, an aide to DC mayor Marion Barry (who supported Solidarity Day), has admitted to withholding his own higher count from the press. Late in the rally, Jordan estimated from the speakers' platform that the throng had surpassed 400,000 – a figure that the aide, who has a reputation for being readily accessible, would not later confirm to reporters. Nine days after the fact, Jordan acknowledged to Zon that he "didn't want to start a controversy" with the Reagan administration; so, fearing possible retaliation in the form of "budget cuts that could hurt the city", he'd allowed more than 140,000 ralliers to drop from the public record. As for the Park Service estimate, Jordan said, "I was very disgusted with it. It was just a political count as far as I was concerned."'

21 Describe a situation that could be represented for each of these algebraic expressions:

(a) $x + 7$

(b) $12 - y$

(c) $6d$

(d) $\frac{2}{3}m$

(e) $3n + 2$

Recreation

22 Write two more lines in each of the following number patterns. Then use variables to write the general equation representing the pattern.

(a) $3 \times \frac{3}{4} = 3 - \frac{3}{4}$

$6 \times \frac{6}{7} = 6 - \frac{6}{7}$

$8 \times \frac{8}{9} = 8 - \frac{8}{9}$

(b) $(5 + \frac{1}{7}) \div \frac{6}{7} = (5 + \frac{1}{7}) + \frac{6}{7}$

$(2 + \frac{1}{4}) \div \frac{3}{4} = (2 + \frac{1}{4}) + \frac{3}{4}$

$(3 + \frac{1}{5}) \div \frac{4}{5} = (3 + \frac{1}{5}) + \frac{4}{5}$

(c) $1 + 2 + 3 + 4 = \frac{4(5)}{2}$

$1 + 2 + 3 + 4 + 5 + 6 + 7 = \frac{7(8)}{2}$

$1 + 2 + 3 + 4 + 5 + 6 + 7 + 8 + 9 + 10 + 11 = \frac{11(12)}{2}$

Projects

23 Read the following poem by Linda Pastan (Robson and Wimp, 1979, p. 50).

(a) Write a brief essay describing what the poem means to you.

(b) Write your own maths poem.

Algebra

I used to solve equations easily.
If train A left Sioux Falls
at nine o'clock, travelling
at a fixed rate,
I knew it would meet train B.
Now I wonder if the trains will crash;
or else I picture naked limbs
through Pullman windows, each
a small vignette of longing.

And I knew X, or thought I did,
shuttled it back and forth
like a poor goat
across the equals sign.
X was the unknown on a motor bike,
those autumn days when leaves flew past
the color of pencil shavings.
Obedient as a genie, it gave me answers
to what I thought were questions.

Unsolved equations later, and winter now,
I know X better than I did.

His is the scarecrow's bitter mouth
sewn shut in cross-stitch;
the footprint of a weasel on snow.
X is the unknown assailant.
X marks the spot
towards which we speed like trains,
at a fixed rate.

REVIEW QUIZ (the meaning of variables) (*Answers at end of chapter*)

1. Write as an algebra equation: 3 less than what number is 17?
2. Translate into words: $14m = 42$.
3. Write an equation which generalizes the arithmetic rule illustrated by the following specific cases:
   $\frac{0}{5} = 0$
   $\frac{0}{12} = 0$
   $\frac{0}{19} = 0$
4. Represent the number of weeks in w years.
5. Using the information in unsolved problem 15, if p equals the population of Oklahoma in 1914, represent, in algebra, the fraction of the population that were dues-paying members of the Socialist Party.

ANSWERS: CHAPTER 19

Solved problems ('you try these')

1 $3 - 3 = 0$
2 3 multiplied by 7 equals 21.
3 15 divided by 15 equals 1.
4 $y = \frac{12}{14 - 8}$ or $y = 12 \div (14 - 8)$. We need the parentheses to indicate that 12 is divided by the answer of $14 - 8$, or 12 is divided by 6.
5 $8 + 5 = m$.
6 3 times some number equals 27, or what number multiplied by 3 equals 27?
7 $n - 0 = n$
8 The area of a triangle is obtained by multiplying $\frac{1}{2}$ by the length of the base and multiplying that answer by the length of the height. Actually $\frac{1}{2}$, b, and h can be multiplied in any order, because whatever order you multiply will give the same answer.
9 $\frac{4}{100}n$ dollars went to indigenous organizations
10 If there were 15 apples and I took n of them, there would be $15 - n$ apples left.
11 $5 \times 7 = 6 \times 6 - 1$; $6 \times 8 = 7 \times 7 - 1$; $n \times (n + 2) = (n + 1) \times (n + 1) - 1$.

Unsolved problems

1(a) $\frac{6}{18} = \frac{1}{3}$
(b) $8 - \frac{1}{2} = 3\frac{1}{2} + 4$
(c) $6(4) = \frac{48}{2}$
(d) $\frac{126.0}{5} = 25.2$

2(a) 3 times 15 equals 45.
(b) 123 divided by 123 is another name for 1.
(c) 4 multiplied by 3 multiplied by 2 is the same as 3 multiplied by 2 multiplied by 4.
(d) 15 divided by 30 equals 0.5.

3(a) $m = \frac{13 + 20}{11}$
(b) $w = 6\ (\frac{12}{4})$
(c) $z = 5(5 + \frac{20}{5})$
(d) $x = \frac{2\ (25 - 10)}{3\ (5 + 10)}$

4(a) $12 + n = 19$
(b) $13 - 6 = y$
(c) $\frac{90}{2} = m$
(d) $5z = 95$

5(a) What number times 3 equals 15?
(b) What number divided by 2 equals 16?
(c) This is the same equation as in (b), but writing $\frac{x}{2}$ as $\frac{1}{2}x$, leads to a translation of what number multiplied by $\frac{1}{2}$ equals 16? Dividing by 2 and multiplying by $\frac{1}{2}$ do the same thing to any number (or $\frac{x}{2} = \frac{1}{2}x$).
(d) What number added to 6 gives the same answer as when that same number is doubled?

8(a) $x(1) = x$, or $1x \times x$, which tells us that x *means* 1 times x.
(b) $\frac{n}{n} = 1$
(c) $x(0) = 0$
(d) $\frac{a}{b} \cdot \frac{b}{a} = 1$

9 $\frac{a}{b} \div \frac{c}{d} = \frac{a}{b} \times \frac{d}{c} = \frac{ad}{bc}$

10 $\frac{a}{b} + \frac{c}{d} = \frac{ad}{bd} + \frac{bc}{bd} = \frac{ad + bc}{bd}$

11(a) The area of a rectangle equals its length multiplied by its width.

(b) The diameter of a circle equals its circumference divided by 3.14. Note that you don't have to understand the things a formula refers to (such as circumference, or π) in order to be able to translate it. You do, of course, have to understand these things in order to be able to use the formula.

12 $10n$ tons of pesticide. The article did not give the actual figures.

13 $\frac{1}{1,000}g$ grants were controversial. For a specific example of how controlled most of these controversial grants are, Mandell relates the following. 'McGeorge Bundy, president of the Ford Foundation, announced on March 16, 1970 that Ford had funded the Southeast Council of LaRaza [a Mexican-American organization] $1.3 million for another two years. This move, according to a *Ramparts* article, was made in order to defuse the more militant demands of Reies Tijerina and his followers in a group called Alianza, who were fighting to reclaim the Tierra Amarilla land grant. The Ford program set up an alternate and less militant group in an attempt to draw off Tijerina's following. On the day Tijerina was sent to prison, Ford announced a $1.5 million grant for a feed-lot in LaJara, Colorado. The ownership of the feed-lot passed into the hands of Claude Lowry, a brother-in-law of Boudinot P. Atterbury, a Ford Foundation official. There is no longer any talk of local Mexican-American ownership of the feed-lot' (p. 82).

14 $\frac{\$370,000,000}{x}$ dollars paid by each customer. The article gives the fact that there are 1.7 million customers, so they will each pay $\frac{370 \text{ million}}{1.7 \text{ million}} \doteq \$218$ over 12 years or about $\frac{218}{12} = \$18$ per year. You can see that each customer wouldn't get that angry – it's only when you see the total picture of how much all the customers will pay for a private business mistake that the outrageousness of the situation becomes clear.

15 $\frac{1,080}{t}$ office holders per municipality. In real life, $t = 340$, so there was an average of $\frac{1,080}{340} = 4$ office holders per municipality who were members of the Socialist Party.

16 $2x$ dollars. This problem is hard because you are given that black income is $\frac{1}{2}$ of white income and here you have to express this relationship from the other side – that white income is twice black income. In 1978 the actual figures were $9,793 for black families and $19,213 for white families.

17 $m - z$ dollars. This problem differs from the others because all quantities are unknown – the only thing known about them is that when you subtract the taxes from the income you get the after-tax income. The actual figures from the article are that their earnings were $2.5 billion and that they paid taxes of about $33 million, a tax rate way below our personal income tax rate.

18 (a) $10d$ cents (because each dime has 10 cents).

(b) $7r$ days

(c) $24y$ hours.

19 p = plastic chair caps; w = washers; s = screwdriver; $2p + 3w + s = 455,000.16$

21 (a)-(d) These of course have numerous answers.

(e) This one is more difficult because first n is multiplied by 3 and then 2 is added to the product. An example of a situation that would lead to this representation is: Lorenzo has 2 more than 3 times as many books as Raphael. If Raphael has n books, represent the number of books Lorenzo has.

22 (a) $n\left(\frac{n}{n+1}\right) = n - \frac{n}{n+1}$

(b) $\left(n + \frac{1}{n+2}\right) \div \frac{n+1}{n+2} = \left(n + \frac{1}{n+2}\right) + \frac{n+1}{n+2}$

(c) $1 + 2 + 3 + \ldots + (n-1) + n = \frac{n(n+1)}{2}$

Review quiz

1 $x - 3 = 17$

2 What number times 14 equals 42?

3 $\frac{0}{x} = 0$

4 $52w$ because each year has 52 weeks.

5 $\frac{12{,}000}{p}$ is the fraction of Oklahoma people who were dues-paying members of the Socialist Party.

Appendix
Critical Mathematics

Often in our highly technological world, an overemphasis on numerical information obscures the reality of people's lives and diverts our attention from ways of changing that reality. For example, disputing which 'side' has exactly how many nuclear weapons, which can destroy how many lives and how much of the environment, hides the more rational, non-numerical issue of why anyone has such weapons. For another example, in a review of Shere Hite's *Women and Love*, Elayne Rapping (1987) makes the important argument that Hite's extensive use of quantitative methods shifts the debate about the effects of systemic sexism to male 'scientific' turf, and thereby obscures the emotional misery of living in a sexist society revealed by the respondents to her surveys. 'Stories are the stuff of literature. They have their own methodology, their own hallowed tradition of legitimacy and honor as a form of truth-telling. They needn't be justified in quantitative, male-defined terms.'

Yet, as this book argues, there are many struggles to humanize our lives that are illuminated, and even driven forward, by numbers. Some major goals of this text have been: to help you develop the confidence to analyse the numbers presented by others about issues of importance to you; to help you develop the knowledge with which to make decisions about the kinds of numerical data you need to analyse those issues; and to help you develop the skills to research and use those statistics.

This appendix concludes with: excerpts illustrating the actual use of statistical data by various social-change organizations; a list of critical publications and/or publishers that analyse social, political and economic issues from statistical viewpoints; a bibliography of references that teach how to research corporations and governmental agencies; and a directory of social-change organizations that make use of numerical data as part of their work.

You might conclude your study of this book by: evaluating the effectiveness of the use of numerical data in each of the excerpts presented; sending for information from the groups whose work is important to you; thinking about how and when other groups might use numbers in their fight for change; and, as a capstone project, collaborating with others to research, produce, and distribute a pamphlet which includes a basic statistical analysis about an issue of mutual concern.

En la lucha,
Marilyn

EDUCATION EXPENDITURE

Excerpts printed below are from *Figuring Out Education Spending: Trends 1978–85 and Their Meaning*, produced by the Radical Statistics Education Group in London. This pamphlet tests the Conservative government's assertions about education against official education statistics and finds that 'public perceptions of a deteriorating service are much nearer the truth than government assertions of improvement'. The Radical Statistics Education Group distributed the pamphlet to those most affected, including parents' and teachers' organizations. They hoped their analysis would be useful to individuals and groups struggling to improve the schools; they also hoped the pamphlet would show that 'anyone with access to an appropriate library and very little expertise could carry out a similar analysis of government spending'. They 'encourage all our readers to continue and expand on our work by looking at new statistics as they are published and by pressing for better statistics to fill many gaps . . . Statistics are often mystifying but they can be empowering, and critical, tools.'

Trends in total expenditure

Figure 1 illustrates the trend in spending on education in the United Kingdom from 1978 to 1985 compared with similar data for defence (i.e. the military) and public order and safety. For each of these three sectors, expenditure in 1980 is fixed at 100. We see that real spending on education remained fairly constant, rising by just 5% between 1979 and 1984 but then falling back by 3% in 1985 to a level lower than in any year since 1979. This contrasts strongly with spending on the military and on public order and safety which rose by 22% and 28% respectively.

Figure 2 presents the percentage of total government expenditure (that is, central and local combined), committed to the military and education from 1978 to 1985 (public order and safety is omitted because it is very much smaller at about 3% of all expenditure). We see that between 1978 and 1983 more was spent on education than on defence but the gap was steadily narrowing, with defence steadily increasing and education steadily falling, and in 1984 and 1985 more was spent on defence than on education.

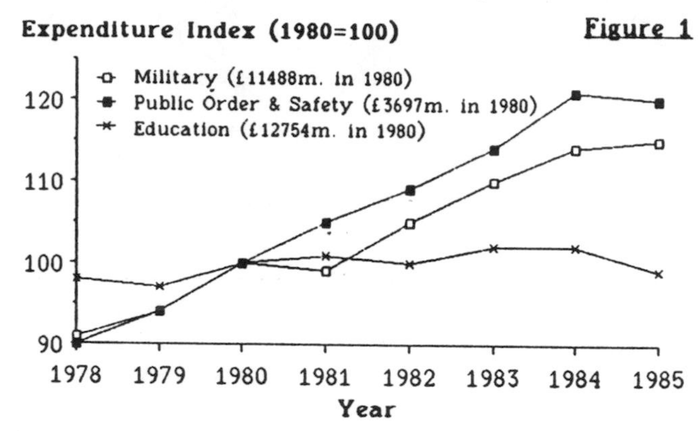

Source: 1986 UK National Accounts

Note: Expenditure reduced by GDP deflator (at market prices), a special type of price index applied to the level of Gross Domestic Product (GDP) to allow for inflation.

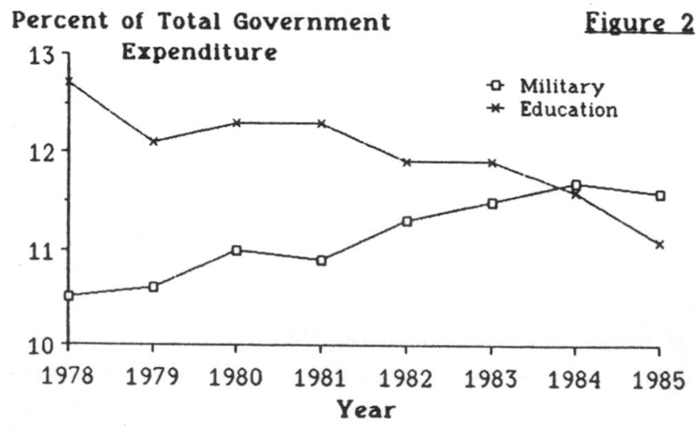

Source: 1986 UK National accounts

A government's priorities are manifested by the way in which it changes expenditure patterns. These graphs give us a clear indication of this government's priorities: the military and public order and safety are more important than education.

WAGES AND PROFITS

Excerpts from *A Report to the Community* by the members of Local 34, AFL-CIO, Yale University's clerical and technical workers (September 1984) describing the conditions under which they were working and explaining that they were intending to strike only because the University rejected their proposal for binding arbitration. The Union produced this booklet to show the community that 'we seek to avoid the division and disruption of a strike'.

Living on a C&T salary

The average full-time Yale clerical and technical employee earns only $13,424 per year. This is the average for all full-timers, including those with very skilled and responsible jobs, and those who have worked at Yale for years.

Most of these employees are women, and many are the sole or principal support of their families.

US Government statistics confirm that this is not a living salary:

1983–84 Average Full-Time Yale C&T Salary	**1981 Cost to Support a Family of 4 in New England***	
$13,424	low standard of living:	$16,402
	moderate standard of living:	$29,213
	high standard of living:	$44,281

*(Fall 1981 family budget costs are the most recent available from the U.S. Bureau of Labor Statistics)

A statistical note:
all statistics in this Report about clerical and technical employees are from data provided by Yale University as of July 30, 1984. Sources for other data are shown. For salary purposes, only full-time clerical and technical employees, who are 84% of the bargaining unit represented by Local 34, are used since comparisons of part-time salaries can be misleading.

Sex and race discrimination

The fundamental economic problem which Local 34 members are determined to address is that we are not paid the real value of the work we do, and are not paid living salaries.

In addition to that basic problem, however, Yale's discrimination is emphasized by the inequalities even within the clerical and technical group. Yale pays clerical and technical women significantly less than men, even though the women have worked at Yale longer.

YALE SALARY DISTRIBUTION BY SEX

Group	Average Salary	No. of Full-Time Employees	Average Years at Yale	Average Years in Present Job
Male	$14,056	367 (18%)	5.6	4.0
Female	$13,290	1724 (82%)	6.0	3.8

Similarly, black clerical and technical employees at Yale are paid significantly less than whites, even though the blacks have worked at Yale longer.

YALE SALARY DISTRIBUTION BY RACE

Group	Average Salary	No. of Full-Time Employees	Average Years at Yale	Average Years in Present Job
White	$13,563	1733 (83%)	5.8	3.7
Black	$12,644	292 (14%)	6.8	4.4

A further breakdown by both race and sex shows that Yale follows the classic pattern of discrimination in America: white men earn the most, followed by white women, black men, and black women.

YALE SALARY DISTRIBUTION BY RACE AND SEX

Group	Average Salary	No. of Full-Time Employees	Average Years at Yale	Average Years in Present Job
White Men	$14,324	293	5.7	4.0
White Women	$13,408	1439	5.8	3.9
Black Men	$12,813	57	5.3	4.0
Black Women	$12,603	235	7.1	4.5

(Note: other racial groups are not listed separately here because no other group includes as much as 2% of the work force.)

When Local 34 first published these very disturbing figures, the Yale administration claimed that they are somehow misleading, while not directly disputing their accuracy. Yale Chaplain John Vannorsdall proposed that the Union and the University agree on an impartial three-person team to verify and publish the facts, so that the community could know the truth.

Local 34 accepted the Chaplain's proposal. The Yale administration said no.

This reluctance on Yale's part is not surprising. The administration has not told the truth when it has stated that the discrimination by race and sex within the clerical and technical group disappears when jobs are analysed by salary grade and years.

Actually, the data show that men earn more than women in 7 of the 11 salary grades covering 76% of the employees; that whites earn more than blacks in 9 of the 11 salary grades covering 80% of the employees; and that these race- and sex-linked discrepancies persist when either years in present job or years at Yale are considered.

Does Yale have the money?

Yale University can afford to end its economic discrimination against its clerical and technical employees. During the 1982–83 fiscal year (the last year for which Yale has published its finances), the University enjoyed extraordinary financial prosperity.

– Yale's endowment increased in value by $351 million, or 46.9%, in one year alone, to a record $1.1 billion.*

– The University had an actual cash surplus for the year of $35 million. In any other corporation, this surplus would be described as net profit.

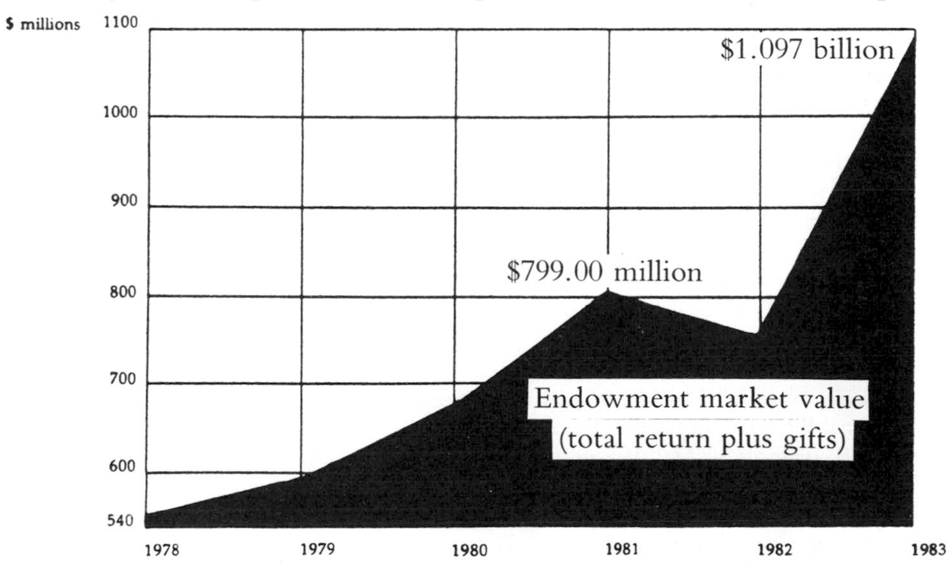

*All data in this section from Yale University's 1982–83 Treasurer's Report.

CRITICAL PUBLICATIONS AND PUBLISHERS

Center for Popular Economics, *The Field Guide to the US Economy*. New York: Pantheon, 1988.

Center for Third World Organising, 3861 Martin Luther King, Jr Way, Oakland, CA 94609. Has published Marcia Henry, *Surviving the United States of America: What You're Entitled to and How to Get It*.

Dollars & Sense (monthly magazine of popular economics), 1 Summer St, Somerville, MA 02143, USA. Tel. 617–628 8411.

Employment Research Associates, 810 Capitol Hill Bldg, 115 West Allegan, Lansing, MI 48933, USA. Tel. 517–485 7655. Booklets include *The Impact of Military Spending on the Machinists Union* and *The Impact of Military Spending on US Congressional Districts*.

Environmental Action (monthly magazine), 1925 New Hampshire Avenue NW, Washington, DC 10036. Tel. 202–745 4870.

Ernest, P., ed., 'The Social Context of Mathematics Teaching', special issue of the journal *Perspectives* 37, April 1988, including articles on mathophobia, politics of numeracy, multicultural and anti-racist maths teaching, gender and maths. Available from School of Education, Univ. of Exeter, Devon EX4 4QJ, UK.

Food Magazine (quarterly), London Food Commission, 88 Old St, London EC1V 9AR. Tel. 01–253 9513.

Gordon, S., and McFadden, D., eds, *Economic Conversion: Revitalizing America's Economy*. Cambridge, MA: Ballinger.

IBON Facts and Figures, IBON Databank, PO Box SM-447, Manila, Philippines.

Interfaith Action for Economic Justice, 110 Maryland Avenue NE, Washington, DC 20002. Has published Robert Greenstein and John Bickerman, *The Effect of the Administration's Budget, Tax and Military Policies on Low-Income Americans*.

John Bibby Books, 1 Straylands Grove, York YO3 OEB. Tel. 0904–424 381. Publishes teaching materials on multicultural maths, women and maths, recreation and humour, investigative maths.

Kidron, M., and Segal, R., *The New State of the World Atlas*, Pan/Pluto, 1984.

Labour Research Department, 78 Blackfriars Rd, London SE1. Tel. 01–928 3649. Publishes monthly magazine, *Labour Research*.

New Internationalist (monthly magazine), 42 Hythe Bridge St, Oxford OX1 2EP. Tel. 0865 728181.

Nutrition Action (monthly magazine), 1501 16th St NW, Washington, DC 20036. Tel. 202–332 9110.

Radical Community Medicine, 55 Fairbridge Rd, London N19 3EW. Tel. 01–281 0922.

Radical Statistics Group, 25 Horsell Rd, London N5 1XL. Booklets include *Figuring Out Education Spending, The Nuclear Numbers Game, The Unofficial Guide to Official Health Statistics, Facing the Figures: What Really is Happening to the NHS?*.

Rose, Stephen J., *The American Profile Poster: Who Owns What, Who Makes How Much, Who Works Where, and Who Lives With Whom*. New York: Pantheon, 1986.

Science for People, 25 Horsell Rd, London N5 1XL. Tel. 01–607 9615.

Science for the People (bimonthly magazine), 897 Main St, Cambridge, MA 02139, USA. Tel. 617–547 0370.

South Africa Perspectives, The Africa Fund, 198 Broadway, NY 10038. Tel. 212–962 1210.

Transnational Information Centre, 9 Poland St, London W1. Tel. 01–734 5902. Booklets include *Working for Big Mac, Beyone the Pale: Transnational Cleaning Companies in Europe, The Other GEC Report, Blood on Their Hands: An Investigation into the Worldwide Activities of BTR, Unilever Monitor*. Available from Collet's International Bookshop, 129 Charing Cross Road, London WC1.

World Priorities, Box 25140, Washington, DC 20007. Has published Ruth Leger Sivard, *World Military and Social Expenditures*.

RESEARCH GUIDES

Corporate Profile Service. Data Center, 464 19th St, Oakland, CA 94612. Tel. 415–835 4692.

How to Do a Hunger Survey. Food Monitor, Institute for Food and Development Policy, 145 Ninth St, San Francisco, CA 94103.

How to Research Your Local War Industry. American Friends Service Committee, 821 Euclid Avenue, Syracuse, NY 13210.

Noyes, D., *Raising Hell: A Citizen's Guide to the Fine Art of Investigation*. Mother Jones, 607 Market St, San Francisco, CA 94105.

Rose, S., *et al.*, *How to Research a Corporation*. Union for Radical Political Economics, 122 West 27th St, NY, NY 10001.

SOCIAL-CHANGE ORGANIZATIONS THAT USE STATISTICS

American Committee on Africa, 198 Broadway, NY 10038. Tel. 212–962 1210.

Center for Defense Information, 1500 Massachusetts Avenue NW, Washington, DC 20005. Tel. 202–862 0700.

Center for Economic Conversion, 222C View St, Mountain View, CA 94041. Tel. 415–968 8798.

Center for Popular Economics, Box 785, Amherst, MA 01004. Tel. 413–545 0743.

Center for Public Advocacy Research, 12 West 37th St, 8th floor, NY 10018. Tel. 212–564 9220.

Center for Science in the Public Interest, 1501 16th St NW, Washington, DC 20036. Tel. 202–332 9110.

Children's Defense Fund, 122 C St NW, Washington, DC 20001. Tel. 202–628 8787.

Coalition for a New Foreign and Military Policy, 712 G St SE, Washington, DC 20003. Tel. 202–546 8400.

Coalition for Basic Human Needs, 54 Essex St, Cambridge, MA 02139. Tel. 617–497 0126.

Committee for Nuclear Responsibility, Box 11207, San Francisco, CA 94101. Nuclear power issues.

Defense Budget Project, Center on Budget and Military Priorities, 236 Massachusetts Avenue NE, Suite 305, Washington, DC 20002. Tel. 202–546 9737.

Environmental Action Foundation, 1525 New Hampshire Avenue NW, Washington, DC 20036. Tel. 202–745 4871.

Institute for Food and Development Policy, 145 Ninth St, San Francisco, CA 94103. Tel. 415–864 8555.

Investor Responsibility Research Center, 1755 Massachusetts Avenue NW, Washington, DC 20036. Tel. 202–939 6500.

Jobs With Peace, 76 Summer St, Boston, MA 02110. Tel. 617–338 5783.

Mobilization for Survival, 853 Broadway, Room 2109, NY, NY 10003. Tel. 212–995 8787.

National Campaign Against Toxic Hazards. 186A South St, Boston, MA 02111. Tel. 617–423 4661.

Nicaragua Statistics Fund, 41 Park Crescent, Bradford BD30 0JZ, UK.

Scientists Against Nuclear Arms (SANA), 9 Poland St, London W1.

US Public Interest Research Group, 215 Pennsylvania Ave SE, Washington, DC 20003. Tel. 202–546 9707.

Women for Economic Justice, 145 Tremont St, Suite 607, Boston, MA 02111. Tel. 617–426 9734.

Women's Health Book Collective, 47 Nicholas Avenue, Watertown, MA 02172. Tel. 924 0271.

Afterword

Ubiratan D'Ambrosio

Mathematics has been increasingly throughout the years an obstacle in school systems. By mystifying mathematics and making it less accessible, society deprives people of a powerful tool for analysing and understanding basic societal issues, for forming their own opinions and political positions. Lack of mathematics understanding may seriously hinder full citizenship.

Critical theory has given both historians of science and educators important insights in the evolution of this process of mystification of mathematics. Let us make it clear that by mathematics we mean the mode of thought which developed in ancient Greece – involving Egyptian, Asian and Arabic contributions – and which was shaped through medieval and Renaissance Europe into its current forms. Mathematics has been an art or technique (*techne* = tics) of understanding, explaining, learning about, coping with, managing, the natural, social and political environment (= *mathema*). This has its origin in the people and to the people it belongs.

Distortions in the history of mathematics tend to obscure its popular origin, create sexist and racist biases and identify mathematical achievement with intellectual privilege. The underlying message is that in order to understand mathematics, and even more to be a mathematician, one must belong to a privileged social category which excludes women, blacks, and working-class people. At the level of public opinion, creating, understanding and using mathematics becomes seen as reserved for a special class of individuals. In the school systems this view is reinforced by a pedagogy oriented towards biased filtering systems, such as tests, quizzes, exams and certifications, all of them carrying the same message that just a few are able to understand mathematics.

Opposing that view, Marilyn Frankenstein offers us an extremely important text. Demolishing this image of 'mathematics for a privileged few', she proceeds to a readable, historically sound and topically relevant discussion of numbers and variables. The first two chapters act organically to overcome students' negative attitudes and to elicit their understanding and involvement.

The same pattern holds through the entire book. By common, everyday examples, mostly taken from current events as reported in periodicals, Frankenstein introduces the reader to the most important topics of mathematics which we encounter in our daily routines. The reader feels immersed in the modern world, developing the ability for a critical reading of newspapers, data and charts. Even when dealing with more advanced topics such as measurements and formulas, Frankenstein keeps the same style, and shows the same capacity for demystifying mathematics; she does ordinary mathematics for ordinary people. This means paving the way for understanding the intricacies of managing one's own economic and political decisions. It is, indeed, a sure way of putting in the people's hands the instruments they need for political emancipation.

Her book brings together relevant current examples with historical perspectives. Well-chosen readings and the several references to the history of concepts allow Marilyn Frankenstein to restore the historicity of socio-cultural bases for the generation, evolution and transmission of mathematics. Without explicitly mentioning ethnomathematics, her book is one of the best comprehensive examples of it. Emphasizing the cultural aspects that we have synthesized into the prefix *ethno* – such as students' experiences, their codes of behaviour and of language – the book shows the evolution of the art or technique of understanding, explaining, learning about, coping with and managing the natural, social and political environment in a way that is very close to students' interests and motivation. In particular, many of these techniques rely on processes like counting, measuring, sorting, ordering, inferring, which throughout history gave rise to the corpus of knowledge which came to be known as mathematics.

Just as reading and writing are more than mere manipulation of letters, syllables and words, so mathematics is more than the ability to calculate. Decisive for the political empowerment of individuals, mathematics, like reading and writing, must be treated as historical, social and political. This book succeeds in bringing the reader to realize these major objectives in an elegant, pleasant way.

Notes

Introduction

1. This problem with making sense of numbers is not confined to 'non-scientists'. A member of the White House science staff in 1970–1, Carl H. Savit, claims that this phenomenon is commonplace at the highest levels of technological research. As an example he discusses a technology-assessment study conducted by a prestigious corporation team of three scientists headed by a senior PhD. A cursory review of their report convinced him that,

'the conclusions . . . were in error by six orders of magnitude. The corporation team . . . reported back that they had redone all their calculations and that the conclusions were correct. I replied that . . . they simply did not make sense . . . It was not until I led the leader of the team step-by-step through the calculations that they discovered they had neglected a factor of 1,000,000 at the head of one column . . . People in engineering and the sciences have a strong tendency to calculate "madly off in all directions" without stopping to think whether their results make numerical sense.' (Savit, 1982)

2. A 1979 French mathematics education study asked a seven-year-old: 'You have ten red pencils in your left pocket and ten blue pencils in your right pocket. How old are you?' The child answered, 'Twenty years old.' When the researcher remarked that the child knew he was not twenty years old, the child replied: 'Yes, but it is your fault: you did not give me the right numbers!' Puchalska and Semadeni believe that:

'Perhaps the most important single reason why students give illogical answers to problems with irrelevant questions or irrelvant data is that those students believe mathematics does not make any sense.'

They go on to argue that the 'social contract' between maths teacher and student does not include the intrustion of the real world. If the question in the above problem had been 'How many pencils do you have?' the student is not supposed to count the real number of pencils in his or her pockets, but is supposed to add and answer 'Twenty pencils'. Consequently, when the question is 'How old are you?' the child may understand that it is not a question about his or her true age, but a part of a 'word problem game' where special rules have to be observed and one's true age has nothing to do with the mathematical reasoning.

I also think that maths 'anxiety' and avoidance are involved – students want to get the maths problem over as soon as possible, with as little thought as possible. Who wants to spend time on activities that are meaningless and have a history of 'failure' attached to them?

3. I have added mathematics to extend the argument concerning the connections between reading and writing made by Anthony Petrosky (p. 20).

4. Even if it were true that unemployment rates were similar when the educational level was the same, that would only shift the question about discrimination to why there is such inequality in educational level.

Chapter 1

1. J. L. Adams argues that the 'inability to incubate' does not 'allow the unconscious to struggle with problems' and that the 'inability to tolerate ambiguity; overriding desire for order; no appetite for chaos' ignores the fact that 'the solution of a complex problem is a messy process . . . One must usually wallow in misleading and ill-fitting data, hazy and difficult-to-test concepts, opinions, values, and other such untidy quantities. In a sense, problem-solving is bringing order to chaos' (p. 55).

Chapter 4

1. Noëlle Bisseret, a sociology researcher in Paris, theorizes that the relationship to time is both a reflection and a constitutive factor in the situation of our society where a small group of people dominate the rest of us.

'The dominants' position gives them hold over the future and enables them to make long-term plans. On the other hand, a system of constraints encloses the dominated in a present from which they cannot take their distance and which compels them to make short-term plans and to make one-way choices. The relationship to time is thus one of possession for some and of privation for the others.' (p. 39)

Chapter 5

1. Petrosky theorizes that 'our comprehension of texts, whether they are literary or not, is more an act of composition – for understanding is composing – than of information retrieval . . .' (p. 19). Rewriting problems in your own words, therefore, is a way truly to understand the given information and the question to be answered.

2. In Volume 2 of this text, I will expand upon the other ways to reduce the load on short-term memory suggested by Horwitz: break down the problem-solving procedure into steps that can be handled one at a time; and develop mental schemata (formulas or visual representations such as graphs) that 'capture certain interrelationships so that they can be held in memory without undue strain' (p. 10).

Chapter 6

1. As Myles Horton, founder of the Highlander Research and Education Center in Tennessee, observed, while meeting with farmers who wanted to know how to do such things as testing their wells for typhoid, 'To my amazement my inability to answer questions didn't bother them . . . You don't have to know the answers! You raise the questions, sharpen the questions, get people to discussing them. We found that in that group of mountain people a lot of the answers were available if they pooled their knowledge' (Thrasher, p. 15).

Chapter 9

1. Zaslavsky goes on to comment that Schmidl's 'views were incompatible with those of the Nazi conquerors of Austria. During the German occupation Dr Schmidl was killed – a victim of Hitler's policy of extermination of the Jews' (p. 15).

2. 'Handicapped' is a word that some groups in the USA want to change to reflect their reality more accurately. Some use 'physically challenged', others use 'differently abled'. As of this writing, to my knowledge, neither of these alternatives is widely used. Which do you prefer and why?

Chapter 18

1. Helena M. Pycior, a history professor at the University of Wisconsin-Milwaukee, recently discovered a spoof-play on Augustus DeMorgan's *Elements of Algebra*, an early (1837) undergraduate algebra text which 'tried to justify the use of negative numbers by "explaining" them as easily rectifiable mistakes . . . In the play, DeMorgan's students cite the principle that the greater may be taken from the lesser as justification of their withdrawal from his classes' (Pycior, 1982).

Chapter 19

1. All the numbers in Al-Khwarizmi's algebra book are expressed as words rather than numerals, for example, 'two and one-half' instead of '$2\frac{1}{2}$', because decimal numerals had been developed in India only a couple of centuries earlier and were not yet used widely. 'Al-Khwarizmi's next great work was a treatise on arithmetic, in which the Hindu Arabic numerals were presented and the place-value system was explained. This textbook was the earliest written on the decimal system' (Arndt, p. 670).

2. Hogben's *Mathematics for the Million* opens with a quote from Tobias Dantzig's *Number: The Language of Science*: 'It is a remarkable fact that the mathematical inventions which have proved to be most accessible to the masses are also those which exercised the greatest influence on the development of pure mathematics.'

References

Adams, J. L. (1974) *Conceptual Blockbusting.* San Francisco, CA: W. H. Freeman.

Anderson, M. (1979) *The Empty Pork Barrel: Unemployment and the Pentagon Budget.* Lansing, MI: Employment Research Associates.

Arndt, A. B. (1983) 'Al-Khwarizmi', *The Mathematics Teacher.*

Ascher, M., and Ascher, R. (1986) 'Ethnomathematics', *History of Science* 24: 125–44.

Babson, S., and Brigham, N. (1978) *What's Happening to Our Jobs?* Somerville, MA: Popular Economics Press.

Beckwith, J. and Durkin, J. (1981) 'Science and the attack on women: girls, boys and math', *Science for the People* September/October: 6–9, 32–5.

Bedell, B. (1982) 'Reaganomics results: 10% jobless', (NY) *Guardian* 20 October.

Bell, M. S. (1972) *Mathematical Uses and Models in our Everyday World.* School Mathematics Study Group.

Bergamin, D. (1971) *Mathematics.* New York: Time-Life.

Berry, J. W. (1985) 'Learning mathematics in a second language; some cross-cultural issues', *For the Learning of Mathematics* 5, 2: 18–23.

Bisseret, N. (1979) *Education, Class Language and Ideology.* London: Routledge & Kegan Paul.

Blumberg, P. (1980) 'Another day, another $3,000: executive salaries in America', in M. Green, *The Big Business Reader.* New York: Pilgrim, pp. 316–32.

Boston Women's Health Book Collective (1976) *Our Bodies, Ourselves.* New York: Simon & Schuster; (1978) London: Penguin.

Bowles, S., and Gintis, H. (1976) *Schooling in Capitalist America.* New York: Basic.

Braverman, H. (1974) *Labor and Monopoly Capital.* New York/London: Monthly Review Press.

Buerk, D. (1981) *Dissertation Abstracts International* 42, 1, 119A: 158–60.

——— (1985) 'The voices of women making meaning in mathematics', *Journal of Education* 167, 3: 59–70.

Burton, L. (1986) *Girls into Mathematics Can Go.* London: Holt Educational.

Buxton, L. (1981) *Do You Panic about Mathematics?* London: Heinemann Educational.

Cockroft Committee (1982) *Mathematics Counts.* London: HMSO.

Cole, M., Gay, J., Glick, J. A., and Sharp, D. W. (1971) *Cultural Context of Learning and Thinking.* New York: Basic.

Commoner, B. (1975) 'How poverty breeds overpopulation', in R. Arditti *et al.*, eds, *Science and Liberation*. Boston: South End Press, 1980, pp. 76–89.

Council on Interracial Books for Children (1982) *Fact Sheet on Institutional Racism*. New York.

Counter-Information Services (1972) *Rio Tinto Zinc Anti-Report*. London.

——— (1978) *Ford Anti-Report*. London.

Desforges, C., and Cockburn, A. (1987) *Understanding the Mathematics Teacher*. London: Falmer Press.

Edelman, M. W. (1980) *Portrait of Inequality: Black and White Children in America*. Washington, DC: Children's Defense Fund.

Edwards, R. C., Reich, M. and Weisskopf, T. E. (1978) *The Capitalist System*. Englewood Cliffs, NJ: Prentice-Hall.

Fadiman, C. (1958) *Mathematical Fantasia*. New York: Simon & Schuster.

Fasheh, M. (1982) 'Mathematics, culture and authority', *For the Learning of Mathematics* 3, 2: 2–8.

Frankenstein, M. (1987) 'Critical mathematics education: an application of Paulo Freire's epistemology', in Ira Shor, ed. *Freire for the Classroom: A Sourcebook for Liberatory Teaching*. New York/London: Heinemann.

Freire, P. (1970a) *Pedagogy of the Oppressed*. New York: Seabury.

——— (1970b) *Cultural Action for Freedom*. Cambridge, MA: Harvard Educational Review.

——— (1985a) *Education for Critical Consciousness*. London: Sheed & Ward.

——— (1985b) *The Politics of Education*. Granby, MA: Bergin & Garvey.

Freire, P., and Macedo, D. (1987) *Literacy: Reading the Word and the World*. Granby, MA: Bergin & Garvey.

Freire, P., and Shor, I. (1987) *Pedagogy for Liberation: Dialogues for Transformation*. London: Macmillan Educational.

Friedberg, R. (1968) *An Adventurer's Guide to Number Theory*. New York: McGraw-Hill.

Gardner, M. (1961) *The 2nd Scientific American Book of Mathematical Puzzles and Diversions*. New York: Simon & Schuster.

Gattegno, C. (1988) *Mathematisation of Awareness*. London: Educational Solutions.

Gerdes, P. (1988) 'On culture, geometrical thinking and mathematics education', *Educational Studies in Mathematics* 19: 137–62.

Gill, D. (1985) 'Political ideology in maths textbooks', *Maths Teaching* 114: 12.

Gordon, D. (1977) *Problems in Political Economics*. Lexington, MA: D. C. Heath.

Gould, S. J. (1981) *The Mismeasure of Man*. New York: Norton.

——— (1985) 'The media isn't the message', *Discover* June: 40–2.

Hemmings, R. (1980) 'Mathematics and multi-ethnic education. Part II: Secondary', *Name Magazine*.

Henderson, D. (1981) 'Three papers', *For the Learning of Mathematics*, 1, 3: 12–15.

Hilton, P. (1980) 'Math anxiety: some suggested causes and cures, Part 1', *Two-Year College Mathematics Journal* 11, 3: 174–88.

Hofstadter, D. (1982) 'Number numbness, or why innumeracy may be just as dangerous as illiteracy', *Scientific American* 245, 5: 20–34.

Hogben, L. (1965) *Mathematics for the Million*. New York: Pocket.

Horwitz, L. (1978) 'A new look at math anxiety', presentation, College of Public and Community Service, Boston, MA.

Hyman, R., and Price, B. (1979) 'Labor statistics', in Irvine, Miles and Evans pp. 222–36.

Irvine, J., Miles, I., and Evans, J., eds, *Demystifying Social Statistics*. London: Pluto.

Joseph, G. G. (1987) 'Foundations of Eurocentrism in mathematics', *Race & Class* 28, 3: 13–28.

Karabel, J. (1972) 'Community colleges and social stratification', *Harvard Educational*

Review 42, 4 November: 521–62.
Kline, M. (1967) *Mathematics for the Liberal Arts*. Reading, MA: Addison-Wesley.
——— (1972) *Mathematical Thought from Ancient to Modern Times*. Oxford/New York: Oxford University Press.
Knill and Fawcett (1982) 'A chilling experience', *The Mathematics Teacher*, April: 320–3.
Kramer, E. E. (1970) *The Nature and Growth of Modern Mathematics*. New York: Hawthorne.
Lappé, F. M. (1975) *Diet for a Small Planet*. New York: Ballantine.
Lappé, F. M., and Collins, J. (1977) *Food First*. New York: Ballantine; (1982) London: Sphere.
Leiner, M. (1981) 'Two decades of educational change in Cuba', *Journal of Reading* 25, 3: 202–14.
Lewontin, R. (1982) 'Are the races different?', *Science for the People* 14, 2: 10–14; reprinted in D. Gill and L. Levidow, *Anti-Racist Science Teaching*, London: Free Association, 1987, pp. 198–207.
Lumpkin, B. (1983) 'Africa in the mainstream of mathematics history', in I. Van Sertima, ed. *Blacks in Science: Ancient and Modern*. New Brunswick, NJ: Transaction, pp. 100–9.
Mandell, B., ed. (1975) *Welfare in America: Controlling the 'Dangerous Classes'*. Englewood Cliffs, NJ: Prentice-Hall.
Manoff, R. K. (1983) 'The week after', *The Nation*, 10 December: 589.
Maxwell, J. (1985) 'Hidden messages', *Mathematics Teaching* 111.
Menninger, K. (1969) *Number Words and Number Symbols: A Cultural History of Numbers*. Cambridge, MA: MIT Press.
McWhirter, N. (1978) *Guinness Book of World Records*. New York: Bantam.
Neft, D. S., Cohen, R. M., and Deutsch, J. A. (1978) *The World Book of Odds*. New York: Grosset & Dunlap.
Ornstein, R. E. (1972) *The Psychology of Consciousness*. San Francisco: W. H. Freeman.
Osen, L. M. (1974) *Women in Mathematics*. Cambridge, MA: MIT Press.
Petrosky, A. (1982) 'From story to essay: reading and writing', *College Composition and Communication* 33, 1: 19–36.
Powell, A. B., and López, J. A. (1989) 'Writing as a vehicle to learn mathematics: a case study', in P. Connolly and T. Vilardi, eds *The Role of Writing in Learning Mathematics and Science*. New York: Teachers College, pp. 269–303.
Profreidt, W. (1980) 'Socialist criticisms of education in the United States: problems and possibilities', *Harvard Educational Review* 50, 4: 467–80.
Puchalska, E., and Semadeni, Z. (1987) 'Children's reactions to verbal arithmetical problems with missing, surplus or contradictory data', *For the Learning of Mathematics* 7, 3: 9–16.
Pycior, H. M. (1982) 'Historical roots of confusion among beginning algebra students: a newly discovered manuscript', *Mathematics Magazine* 55, 3: 150–6.
Rapping, E. (1987) 'Women's pain obscured by Hite's science', (NY) *Guardian* 30 December.
Reed, D. (1981) *Education for Building a People's Movement*. Boston, MA: South End Press.
Robson, E., and Wimp, J. (1979) *Against Infinity: An Anthology of Contemporary Mathematical Poetry*. Parker Ford, PA: Primary Press.
Rogers, J. (1984) 'The politics of voter registration', *The Nation* 21 July: 47–51.
Rosamund, F. (n.d.) 'Mathematics: the female voice'. Unpublished MS.
Savit, C. H. (1982) Letter to the editor, *Scientific American* 247, 1: 6.
Seldon, S. (1983) 'Biological determinism and the ideological roots of student classification', *Journal of Education* 167, 2: 175–91.
Shaw, M., and Miles, I. (1979) 'The social roots of statistical knowledge', in Irvine, Miles and Evans, pp. 27–38.

Shor, I. (1978) 'The working class goes to college', in T. M. Norton and B. Ollman, *Studies in Socialist Pedagogy*, New York: Monthly Review Press, pp. 107–27.
Singh, E. (1985) *Mathematics Teaching – Culture and Racism*. London: ILEA Centre for Anti-Racist Education.
Stack, C. B., and Semmell, H. (1975) 'Social insecurity: welfare policy and the structure of poor families', in Mandell, pp. 89–103.
Statistical Abstracts of the United States 1986. Washington, DC: US Department of Commerce, Bureau of the Census.
Stempien, M., and Borasi, R. (1985) 'Students writing in mathematics: some ideas and experiences', *For the Learning of Mathematics* 5, 3: 14–17.
Stone, C. (1975) 'Let's abolish the SAT, IQ & ETS, too', *The Black Collegian*.
Thrasher, S. (1982) 'Fifty years with Highlander', *Southern Changes*.
Tobias, S. (1978) 'Managing maths anxiety: a new look at an old problem', *Children Today* September/October: 7–9, 36.
Waldon, R., and Walkerdine, V. (1982, 1985) *Girls and Mathematics*, parts 1 and 2. London: Institute of Education.
Walkerdine, V. (1988) *The Mastery of Reason*. London: Routledge.
Weissglass, J. (1976) 'Small groups: an alternative to the lecture method', *Two-year College Mathematics Journal* 7: 15–20.
——— (1979) *Exploring Elementary Mathematics*. San Francisco: W. H. Freeman.
Williams, G. (1971) *African Designs from Traditional Sources*. Mineola, NY: Dover.
Zaslavsky, C. (1973) *Africa Counts*. Boston, MA: Prindle, Weber & Schmidt.
——— (1983) 'Essay review of *Black Mathematicians and Their Works* and *A Negro History Compendium*', *Historia mathematica* 10: 105–15.
Zinn, H. (1980) *A People's History of the United States*. New York: Harper & Row.

Periodicals addresses

International Study Group on Ethnomathematics Newsletter, c/o Gloria Gilmer, 2001 W. Ulnet St, Milwaukee, WI 53205. Tel. 414–933 2322.
For the Learning of Mathematics, 4336 Marcil Avenue, Montreal, Quebec H4A 2Z8.
Journal of Education, School of Education, Boston University, 605 Commonwealth Avenue, Boston, MA. 02215.

This first edition of
Relearning Mathematics: A Different Third R – Radical Math(s)
was finished in August 1989.
It was set in 10/13 Bembo Roman on a Linotron 202
and printed on a Miller TP41,
onto 80gm wood-free paper.

The book was commissioned and edited
by Les Levidow,
copy-edited by Peter Phillips,
designed by Wendy Millichap,
and produced by Simona Sideri and Martin Klopstock
for Free Association Books.

Bakuba Strip Patterns ***(printed on the ornamental endpapers)***
How might mathematics be relevant for combating racial and colonial prejudice? Paulus Gerdes, a mathematics teacher at Eduardo Mondlane University in Mozambique, argues that

> A cultural-mathematical reaffirmation plays a part. It is necessary to encourage an understanding that our peoples have been capable of developing mathematics in the past, and therefore – regaining cultural confidence – will be able to assimilate and develop the mathematics we need. Mathematics does not come from outside our African, Asian and American-Indian cultures. (p. 140)

For example, by studying the geometry 'culturally frozen' in Mozambican square-woven buttons, and by analysing the patterns in that form, maths education students realize that 'Had Pythagoras – or somebody else before him – not discovered this theorem, *we* would have discovered it' (p. 151). This cultural-mathematical confidence leads students to question whether their ancestors actually did discover this theorem; and, if so, why they don't know this history. 'By "defrosting frozen mathematical thinking" one stimulates a reflection on the impact of colonialism, on the historical and political dimensions of mathematics [education]' (p. 152). This 'defrosting' is one aspect of a new focus in mathematics education, 'ethno-mathematics'.

The Bakuba strip patterns from Zaire that decorate the end papers of this text are another example of the mathematical knowledge inherent in material culture. They represent the seven symmetrically different border patterns that can be created from a unit design. Through group theory we can show that using a unit design, with various combinations of the symmetry operations of translations and reflections, all border patterns will have one of these seven forms: no others are possible. The people who created these designs may not have used formal group theory but they clearly explored and discovered all the possibilities of symmetry operations on strip patterns.

As you work through this textbook, I hope you will try to 'defrost' your hidden, frozen mathematical knowledge and use it as a bridge to connect whatever new mathematics you learn here with the mathematics you already know.

Reference: Paulus Gerdes (1988) 'On culture, geometrical thinking and mathematics education', *Educational Studies in Mathematics* 19: 137–62.